DEVELOPING
BIOLOGICAL LITERACY

A Guide to Developing Secondary and Post-secondary Biology Curricula

D1315889

BSCS

KENDALL/HUNT PUBLISHING COMPANY
4050 Westmark Drive   Dubuque, Iowa 52002

Photo Credits

Cover Photos

Left photo taken by Gordon Uno, BSCS staff
Right photo taken by Ripon Microslides, Inc., PO Box 262, Ripon WI 54971

Interior Photos

Gordon Uno, BSCS staff, page 6
John A. Moore, University of California, Riverside CA, page 112

All other photos were taken by Sigurd Olsen and
reproduced by permission from Ingrith Deyrup-Olsen.

For information about content, please contact:
BSCS
Pikes Peak Research Park
5415 Mark Dabling Blvd.
Colorado Springs CO 80918-3842

This material is based upon work supported by the National Science
Foundation under Grant No. USE-9150808. Any opinions, findings, and
conclusions or recommendations expressed in this publication are those
of the authors and do not necessarily reflect the views of the National
Science Foundation.

Additional Reviewers

Beth Brooks, Colorado Springs, CO
Edward Drexler, Pius High School, Milwaukee, WI
Danine L. Ezell, Bell Junior High School, San Diego, CA
Joseph L. Graves, University of California, Irvine, CA
Toni Miller, Springfield High School, Akron, OH
Harold Pratt, Jefferson County Public Schools, Golden, CO

Contributing Essayists

Francisco J. Ayala, University of California, Irvine, CA
John Tyler Bonner, Princeton University, Princeton, NJ
Christian de Duve, Rockefeller University, New York, NY
Franklin M. Harold, Colorado State University, Fort Collins, CO
Paul DeHart Hurd, Stanford University, Palo Alto, CA
James L. Gould, Princeton University, Princeton, NJ
Lynn Margulis, University of Massachusetts, Amherst, MA
Ernst Mayr, Harvard University, Cambridge, MA
John A. Moore, University of California, Riverside, CA
Knut Schmidt-Nielsen, Duke University, Durham, NC
Eugene P. Odum, University of Georgia, Athens, GA
Dorion Sagan, Sciencewriters, Amherst, MA

Table of Contents

Executive Summary

In biology courses at the college and precollege levels, the struggle to improve science education is, in part, the result of the diverse goals set by American educators. With support from the National Science Foundation (NSF), the Biological Sciences Curriculum Study (BSCS) presents recommendations that bring continuity and cohesion to the myriad goals of biology education.

Based on the efforts of forty-one scientists and science educators, the project's final report, titled Developing Biological Literacy, is a guide to designing biology curricula at the secondary and post-secondary levels. The guide is not a curriculum, rather it is a curriculum framework from which individuals may choose various recommendations to organize their program. We have synthesized the contributions of contemporary reports into a curriculum framework that is more specific than any report, yet this guide does not represent a fully developed program. The curriculum framework presented here is an intermediate step between the general recommendations of other reports and the specific details of programs developed by individuals or organizations. This guide includes background information and specific suggestions that local school districts, colleges, universities, or national groups can use as the basis for developing and implementing new biology programs. Each of the recommendations in this summary is elaborated within the main text of the guide.

Major Recommendations

In this guide we make three major recommendations: the content of biology must be unified by the theory of evolution; biology classes must provide opportunities for students to experience science as a process and to understand science as a way of knowing; and programs should help students develop biological literacy. These recommendations apply to high schools, community colleges, and four-year colleges and are provided in this guide for individuals and organizations to use in developing their programs.

Goals for a Biology Program

The objective of developing biological literacy must be translated into specific goals for biology education. Education in biology should sustain students' interest in the natural world, help students explore new areas of interest, improve their explanations of biological concepts, help them develop an understanding and use of inquiry and technology, and contribute to their making informed personal and social decisions. More specific goals are embodied in answers to the question: "What should biology education achieve?" This guide specifically addresses the following challenges for biology education:

- increasing concentration on major unifying principles of biology and biological inquiry;
- teaching biology in contexts that have personal, social, and ethical meaning,
- developing active learning environments through better use of discussions, laboratories, field experiences, and educational technologies,
- providing implementation strategies for new programs, and
- improving the biological literacy of all students.

Goals for Biology Students

What should a biologically literate student know, value, and be able to do after completing a contemporary biology program? A biologically literate individual should understand the unifying principles and major concepts of biology, the impact of humans on the biosphere, the processes of scientific inquiry, and the historical development of biological concepts. In addition, a biologically literate individual should develop appropriate personal values about scientific investigations, biodiversity and cultural diversity, the impact of biology and biotechnology on society, and the importance of biology to the individual. Finally, a biologically literate individual should be able to think creatively and formulate questions about nature, reason logically and critically and evaluate information, use technologies appropriately, make personal and ethical decisions related to biological issues, and apply knowledge to solve real-world problems.

The implication from the list above is that if students are to develop biological literacy, all programs should present the biological knowledge, skills, and values necessary for them to become effective citizens in society. The contemporary phrase "science for all" implies that any new biology program should address the needs and concerns of *all* students, which, by definition, includes those who traditionally have been underrepresented in the sciences.

Biological Literacy

This BSCS guide recognizes that biological literacy extends beyond memorizing the definitions of scientific terms. The development of biological literacy is a lifelong, continuous endeavor. People do not attain literacy in biology or any other subject and then stop learning; rather, we continue building upon the knowledge, understanding, and skills we acquire through time. Thus, we hope that students will use the knowledge and skills they obtain in biology classes to develop their literacy throughout their lives. We propose a model that includes four distinct levels of biological literacy: Nominal, Functional, Structural, and Multidimensional.

Students often come to a class with a *nominal* level of literacy—that is, they are literate "in name only." Students may recognize biological terms as being related to natural phenomena, but they cannot provide scientifically valid explanations of the phenomena and may express misconceptions. At the *functional* level of literacy, students can define terms correctly, but that ability is based on the memorization of information with little under-standing. Students are *structurally* literate if they have constructed appro-priate explanations based on their experiences within class and can discuss and explain concepts in their own terms. At the *multidimensional* level, students can apply the knowledge they have gained and the skills they have developed to solve real-world problems that may require the integration of information from other disciplines such as sociology, economics, and political science. We recommend that curriculum developers and biology instructors provide opportunities for students that help them reach the structural and multidimensional levels of biological literacy. Figure A illus-trates the model of biological literacy.

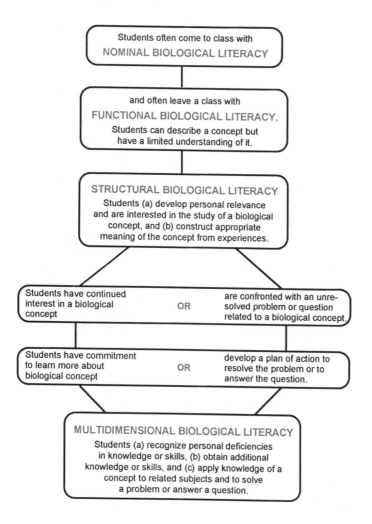

Figure A: Biological literacy model

Developing Biological Literacy in Students

Structural and multidimensional biological literacy are appropriate levels of literacy for all students. Structural literacy requires an inquiry-based biology program where students construct their own, scientifically valid explanations of biological structures, concepts, and natural phenomena. Structural literacy, of course, also requires student interest, and instructors can increase student interest in biology by using hands-on, exciting laboratories and activities, and by helping students see the relevance of the subject to their own lives. For example, to engage students in a learning activity, a local land-use issue, such as where to put a new landfill, can demonstrate that concepts learned in the classroom (e.g., population growth, resource use, and carrying capacity) have relevance and practical application in the real world.

One can help students develop multidimensional literacy by several different routes but, as with structural literacy, personal interest is one key. For instance, if a student or a close friend or relative contracts a disease, that student may become engaged in the study of the biology of that disease. For those students who are curious about any subject, an instructor can point to additional readings and research that addresses unanswered questions and that might lead to open-ended, independent investigations.

Other students may continue study of a biological concept if an instructor presents a problem that affects them directly or if the students find themselves confronted with personally relevant local, regional, or global issues. If necessary, instructors can initiate this process by posing problems or identifying biological issues within the community that stimulate student interest. Students then can be challenged to apply their knowledge to the problem's resolution. An unresolved local public issue also gives students the opportunity to become directly involved in a multidimensional investigation that is relevant, in which they practice the processes of scientific investigation, *and* in which they learn biological concepts and the connections to other concepts and disciplines. An issue-centered or problem-based approach to teaching as described above organizes the curriculum on opportunities for students to practice and develop critical thinking, analysis, and decision-making skills. This BSCS guide outlines other instructional strategies that may help students develop structural and multidimensional literacy.

Instructional Strategies

The activities of a biology class make up the instructional strategies, and we recommend that biology educators consider varying the activities they use. The use of a variety of teaching strategies accommodates many student learning styles and provides opportunities that will help different students construct their understanding of biological concepts. Inquiry-oriented activities should be central to biology education because they move students toward greater independence as individual thinkers and investigators. Table A describes a variety of activities that might be used in a biology program at different levels of education and suggests the appropriate use for each activity.

Instructors should move from a lecture-only format. The greatest change in a lecture course comes from the addition of a laboratory or other investigative experience and this guide recommends that laboratory and field experiences constitute the central components of any program. Finally, biology programs should incorporate educational technologies in ways that are appropriate to biological inquiry, that are adequate to program requirements, and that accommodate diverse learning styles.

LEARNING/TEACHING STRATEGIES	High School	2-Year College	4-Year College
Student-centered Activities			
Direct Experience Activities	✓✓✓	✓✓✓	✓✓✓
• Laboratories	✓✓✓	✓✓✓	✓✓✓
• Field Experiences	✓✓✓	✓✓	✓✓
• Inquiry Activities	✓✓✓	✓✓✓	✓✓✓
• Individual Research Projects	✓✓	✓✓	✓✓✓
• Issue-Centered Problems	✓✓✓	✓✓✓	✓✓✓
Group Activities	✓✓✓	✓✓	✓✓
• Discussion	✓✓✓	✓✓✓	✓✓✓
• Debates	✓✓	✓✓	✓✓
• Cooperative Learning	✓✓✓	✓✓	✓✓
Individual Activities	✓✓✓	✓✓✓	✓✓✓
• Student Writing	✓✓✓	✓✓✓	✓✓✓
• Student Reading	✓✓✓	✓✓✓	✓✓✓
• Student Speaking	✓✓	✓✓	✓✓
• Student Explanations of Concepts	✓✓✓	✓✓✓	✓✓✓
• Analysis of Data	✓✓✓	✓✓✓	✓✓✓
Technology Related Activities	✓✓	✓✓	✓✓
• Audiovisuals	✓✓	✓✓	✓✓
• Computer Simulations	✓✓	✓✓	✓✓
• Kits and Manipulatives	✓✓	✓✓	✓
Teacher-centered Activities			
Lectures	✓	✓	✓
Demonstrations	✓✓	✓✓	✓✓
✓✓✓ = ESSENTIAL	✓✓ = OPTIONAL		✓ = NONESSENTIAL

Table A: Learning/teaching strategies with suggested emphases for secondary and post-secondary programs

An Instructional Model

In recent years, cognitive psychologists have shed some light on the learning process and proposed a model that may accommodate all students. The model is referred to as *constructivism*, a term that expresses a dynamic and interactive conception of human learning. Students redefine, reorganize, elaborate, and change their initial concepts through interaction with their environment, including other individuals. This guide recommends the use of an instructional model and includes the following as a model that might be used in the design of a biology program. This instructional model consists of five phases: the **engagement** phase initiates learning, making connections between past and present learning experiences, and organizing students' thinking toward the learning outcomes of current activities. The **exploration** phase of the teaching model provides students with a common base of experience within which current concepts, processes, and skills are identified and developed. The **explanation** phase focuses students' attention on particular aspects of their engagement and exploration experiences and provides opportunities to demonstrate their conceptual understanding, process skills, or behaviors. The **elaboration** phase challenges and extends the students' conceptual understanding and skills. Finally, the **evaluation** phase encourages students to assess their understanding and abilities and provides opportunities for teachers to evaluate student progress toward achieving the educational objectives.

Asking the Right Questions

To investigate any natural phenomenon or problem, students need to ask the right questions. There are a few, general questions that one can ask to learn more about any aspect of biology, and mastery of these questions can empower students to pursue investigations on their own and to develop higher levels of biological literacy throughout their lives. **Description** questions such as "Is this typical?" should be asked when a natural phenomenon or problem is first encountered. **Investigation** questions such as "What causes it to happen?" will help students explore the phenomenon or problem, and **Prediction** questions such as "What are the future risks, benefits, or consequences?" should help students think about the future. Finally, **Evidence** questions such as "Are there any other explanations?" allow students to assess the validity of information used to explain the phenomenon or problem.

Curriculum Themes

A curriculum theme is an instructional orientation that helps organize and deliver a biology program to students. It serves as a paradigm or frame of reference to help organize the structure of the biology *program*. For example, with evolution as a *curriculum theme*, an instructor may choose laboratories that deal with mutation, genetic variation, and natural selection, and discussions that focus on speciation and extinction. The use of curriculum themes should help students understand and appreciate the interrelatedness of science and the interactive nature of the scientific

CURRICULUM THEMES	High School	2-Year College	4-Year College
Nature of Science/Science as a Process	✓✓✓	✓✓✓	✓✓✓
Unifying Principles	✓✓✓	✓✓✓	✓✓✓
Biological Knowledge	✓✓	✓✓	✓✓
Personal Aspects of Biology	✓✓✓	✓✓	✓
Social Aspects of Biology	✓✓	✓✓	✓✓
Ethical Aspects of Biology	✓✓	✓✓✓	✓✓✓
Technology	✓✓	✓✓	✓✓
Historical Development of Biological Ideas	✓✓	✓✓	✓
✓✓✓ = ESSENTIAL ✓✓ = OPTIONAL ✓ = NONESSENTIAL			

Table B: Curriculum themes with suggested emphases for secondary and post-secondary programs

investigative process.   Table B presents several curriculum themes and indicates which might be most appropriate at different levels of education.

Assessment Strategies

Assessment instruments and instructional strategies are closely related to each other and to the biological content of a course.   Assessment, therefore, should be one of the first considerations in the design of a biology curriculum.   Assessment should be varied, reflect the diverse learning and teaching strategies represented in the biology program, and outline for students the expectations for learning science.   Instructors must assess those aspects of the program they say they value.

One can assess diverse students in a classroom by using diverse methods, including nontraditional assessment instruments such as portfolios and student productions of simulations and models.   Assessment instruments are most valuable when they focus on higher-order thinking skills, understanding and use of biological knowledge, and the demonstration of competence in a setting that is meaningful to students.   Assessment should be an ongoing process that begins with a determination of the information and skills students bring to the class and continues with documentation of their progress throughout the course.   Table C provides recommendations for the use of different assessment strategies; the emphasis varies across high school, two-year, and four-year colleges.

ASSESSMENT	High School	2-Year College	4-Year College
Written Examinations			
Multiple Choice Questions	✓✓	✓✓	✓✓
Essay Questions	✓✓	✓✓✓	✓✓✓
Constructed Response Questions	✓✓✓	✓✓	✓✓
Extended Written Work			
Research Papers	✓	✓	✓✓
Project and Laboratory Reports	✓✓	✓✓	✓✓✓
Self Assessment and Journals	✓✓	✓	✓
Student Activity Assessment			
Performance Assessment	✓✓✓	✓✓	✓✓
Laboratory Practicals	✓✓	✓✓	✓✓
Oral Reports/Oral Examinations	✓✓	✓	✓
Interviews with Students	✓	✓	✓
Research Projects	✓✓	✓✓	✓✓✓
Observations of Students	✓✓	✓	✓
Peer Group Assessment	✓✓	✓	✓
Multidimensional Evaluation			
Construction of Models and Simulations	✓✓	✓✓	✓✓
Portfolios	✓✓	✓	✓
✓✓✓ = ESSENTIAL ✓✓ = OPTIONAL ✓ = NONESSENTIAL			

Table C: Assessment strategies with suggested emphases for secondary and post-secondary programs

Unifying Principles of Biology

Biologists study living systems. There are millions of different species of organisms on Earth today, and one goal of biologists is to understand how these species differ from one another. Living systems, however, also have many characteristics in common, and the principles on which biology is founded apply to all organisms. The unifying principles help to organize the biological content of a course and the patterns and processes of natural phenomena in the living world, and they should play a central role in the design of any program of study. For example, when *Evolution: Patterns and Products of Change* is used as a unifying principle of biology, cell parts, tissues, and hormones may be seen as adaptations that have contributed to an organism's successful reproduction and to a species' reproductive success through evolutionary history.

The unifying principles outlined in this guide represent a comprehensive foundation for the biological sciences and thus are common to high school, two-year, and four-year college programs. Individual programs may emphasize a molecular, organismic, or ecological level of organization, and they may give greater emphasis to inquiry or societal problems, but the unifying principles should provide the foundation for the biological knowledge in all contemporary biology programs. This guide outlines <u>six unifying principles</u> that organize biological concepts and that we suggest educators use in biology courses to help unify the otherwise disparate facts of biology. These principles are: **1. Evolution: Patterns and Products of Change,** which reflects that living systems change through time; **2. Interaction and Interdependence, which reflects that living systems interact with their environment and are interdependent with other systems; 3. Genetic Continuity and Reproduction,** which reflects that living systems are related to other generations by genetic material passed on through reproduction; **4. Growth, Development, and Differentiation,** which reflects that living systems grow, develop, and differentiate during their lifetimes based on a genetic plan that is influenced by the environment; **5. Energy, Matter, and Organization,** which reflects that living systems are complex and highly organized, and that they require matter and energy to maintain this organization; and **6. Maintenance of a Dynamic Equilibrium,** which reflects that living systems maintain a relatively stable internal environment through their regulatory mechanisms and behavior.

Major Biological Concepts

Members of the project team identified major biological concepts to serve as a foundation for a core curriculum in biology; lists of these appear in the body of this guide. Based on these lists, this guide identifies twenty major concepts in the context of associated unifying principles that all students at the high school and college levels should understand at the end of their biology program. The twenty major biological concepts and their related unifying principles are:

Evolution
1. The patterns and products of evolution, including genetic variation and natural selection
2. Extinction
3. Conservation biology, including wise use of resources
4. Characteristics shared by all living systems
5. Overview of biodiversity, including specialization and adaptation demonstrated by living systems

Interaction and Interdependence
6. Environmental factors and their effects on living systems
7. Carrying capacity, and limiting factors
8. Community structure, including food webs and their constituents
9. Interactions among living systems
10. Ecosystems, nutrient cycles, and energy flow
11. The biosphere and how humans affect it

Genetic Continuity and Reproduction
12. Genes and DNA, and the effect of interactions between genes and the environment on growth and development
13. Patterns of inheritance demonstrated in living systems
14. Patterns of sexual reproduction in living systems

Growth, Development, and Differentiation:
15. Patterns of development
16. Form and function

Energy, Matter, and Organization:
17. Hierarchy of organization in living systems
18. Metabolism, including enzymes and energy transformation

Maintenance of a Dynamic Equilibrium:
19. Homeostasis, the importance of feedback mechanisms, and certain behaviors
20. Human health and disease

Developing a Contemporary Biology Program

We suggest that you begin designing your biology program by asking yourself these three questions: (1) What are the goals of the program for my students and me? (2) How do I intend to reach these goals? and (3) How will I know I have achieved these goals? To establish appropriate goals, we return to the question, "What do you want your students to know, value, and be able to do after completing your biology program?" To help students develop structural or multidimensional levels of biological literacy, we recommend the following:

- Students should experience science as a process. This means that each program should include investigative experiences for all students.
- Instructors should focus on a limited number of major biological concepts that are linked by the unifying principles of biology.
- Programs should include varied instructional strategies and appropriate assessment procedures that challenge students' higher-order thinking and reasoning skills and their ability to use the processes of science and to solve problems.
- Students should have ample opportunities for discussion, explanation, and linkage of biological principles and concepts.
- Information must be relevant to the personal lives of students and provide opportunities for students to apply this knowledge in the resolution of real-world problems.

Table D is a form that can serve as a model of how you might evaluate your current biology program, set goals, and reach those goals.

Program Component	Current Practices	Long-term Changes in Practices	How Changes Will Be Made—Short-term and Long-term Actions
Goals for Students and Programs (See Chapters 3 & 4)			
Unifying Principles (See Chapter 5)			
Biological Knowledge (See Chapter 5)			
Curriculum Themes (See Chapter 4)			
Instructional Strategies			
Lecture			
Laboratory			
Educational Technology			
Assessment Strategies (See Chapter 4)			
Developing Biological Literacy (See Chapter 3)			
Personal Needs of Students			

Table D: Example of how to evaluate your biology program (See Figure 6-1)

A *Final Perspective on Biology Education*

An effective biology education combines *what* is taught about the biological sciences and *how* students develop biological ideas and skills. At the end of this guide, we have included a cube that can serve as a reminder of the major components of a contemporary biology program. On this cube, which illustrates Biology Education, biological information is represented by two sides, *Unifying Principles of Biology* and *Biological Knowledge*, and is shown to be unified by the theory of evolution. Biology is a scientific discipline, and this is illustrated on a third side of the cube—*Biology is an Active Process*, outlining components of a scientific process of investigation. Two of the remaining sides, *Curriculum Themes* and *Learning/Teaching/Assessment Strategies*, represent what instructors emphasize in their biology courses and how students learn about biology. The sixth side, *Biological Literacy*, incorporates what students should know, value, and be able to do at the end of their experience in a biology program.

We encourage you to assemble the cube, place it on your desk, and consider what might be part of your curriculum in biology.

(Note: the BSCS copyright goes on opposite sides of the cube.)

Developing Biological Literacy

An Introduction

Beginning with A Nation at Risk in April 1983, a proliferation of reports proclaimed the need to improve American education and provided myriad recommendations promoting that goal. The number of reports is a credit to the importance Americans attach to education, but the variety of recommendations is overwhelming and sometimes contradictory. Recommendations such as those calling for increases in required courses and the length of the school day and school year were implemented first because they were easiest. The most difficult aspects of educational reform remain: improving the quality and appropriateness of curriculum, instruction, and assessment. Especially critical is the implementation of changes in science classrooms.

The 1990s, therefore, must be devoted to educational reform through action, and educators must implement changes that are unique to different educational levels (K-16) and to specific disciplines such as biology. Educators also must provide concrete models for those who wish to reform their programs. This report by the Biological Sciences Curriculum Study (BSCS) outlines a curriculum framework that can be used to guide improvement of biology programs in high schools, community colleges, four-year colleges, and universities. The guide also provides a substantive structure for development of materials by nonprofit organizations, textbook writers, and commercial publishers.

In the early 1990s "scientific literacy for all citizens" emerged as the objective for reform of science curricula, instruction, and assessment. Although scientific literacy identifies broad and comprehensive changes in science education, it also expresses an emphasis on general education, an orientation that distinguishes this reform from others in recent history. The 1992 Public Health Service report *Prologue to Action: Life Science Education and Science Literacy* addresses scientific literacy:

Although the United States is in the midst of a life science revolution, most Americans are ill equipped to understand the fundamental scientific concepts that underlie that revolution or to make sound judgments about the implications of scientific and technological progress. Scientifically literate people should be able to participate in discussions of contemporary scientific issues, apply scientific information in personal decision making, and locate valid scientific information when they need it. These components of scientific literacy in turn require that people be able to understand and evaluate publicly disseminated information on science and technology, distinguish science from pseudoscience, and interpret graphic displays of scientific information. Because the rapid pace and complex nature of advances in the life sciences raise difficult issues of ethics and public policy for the Nation, the public's ability to understand science and technology is critical to the Nation's future.

In the essay that concludes this chapter, Paul DeHart Hurd describes many features of the biological revolution and identifies the issues that connect the life sciences to all individuals in society. For most biologists and biology educators, the importance of biology in society is beyond question. Most people, however, have only a vague understanding of the biological principles and processes of inquiry essential to their role as citizens.

The objective of scientific literacy for all citizens challenges biology educators to design programs that maintain the integrity of the discipline while providing an emphasis that places biology in a personal, social, and ethical context. In this report, the BSCS focuses on one major component of scientific literacy, biological literacy, by providing a guide to *Developing Biological Literacy*.

ESSAY

Biology in Transition

by Paul DeHart Hurd

The modern image of biology differs vastly from that of the last century. These shifts are characterized by changes in the conceptual structure of biological knowledge, a reorganization of the discipline, and how research is done. This short essay portrays characteristics of modern biology, their meaning within culture, and their significance for the education of the young.

Since the turn of this century, biology as a research discipline has been fractionated into thousands of research fields at an exponential rate. No one knows the precise number of research fields that characterize today's biology beyond the fact that 20,000 journals exist to report findings, a third of these new since 1979. Each of these fields has its own language, conceptual structure, inquiry processes, and research concerns. Boundaries separating one field from another are fuzzy and have meaning only to the investigator. Although biology is a singular noun, it no longer has meaning as a discipline, but is simply a way of classifying thousands of diverse research fields.

Since midcentury, biological research has become increasingly hybridized into fields such as biophysics, biochemistry, biogeochemistry, biotechnology, and genetic and clinical engineering. Not only is biology integrated with other sciences but also with social, behavioral, and philosophical fields, for example, sociobiology, psychobiology, human ecology, the cognitive sciences, and bioethics. The frontiers of biological research are found at the nodes of these interdisciplinary and cross-disciplinary sciences.

The nature of research in the biological sciences extends beyond the inductive laboratory experiment and controlled observation described in school textbooks. Most of biological research today is characterized by a search for interactions, relationships, or connections of biological systems, multideterminations of problems, and synthesis. These holistic approaches recognize that life can be understood in the context of unity or an ecological matrix.

The complexities of investigating biological systems and the amount of specialized knowledge required to advance the study of a biological problem have led to the conduct of research by teams, sometimes a coordination of international teams as in the research on AIDS and the human genome project. Depending on the problem, there is a mix of specialists from related fields but with diverse insights, instrumentation specialists, and that indispensable electronic "person," the computer. Computers serve to feed up-to-date information to the researchers, to record, organize, and produce possible models, to interpret new data, or, with the use of parallel computers, to join theory with experiment and provide "virtual reality."

As the sciences have become a central part of our culture and a major force in social and economic progress, they have become socialized. As a result, research in modern biology has become more socially oriented than theory driven. Further evidence of this shift is found in the commercialization of research. More than half of all biologists are employed in industries such as biotechnology, pharmacology, and agriculture. Under these conditions, investigations may be controlled by ethical (e.g., use of fetal tissue and animals for experimental studies); moral (e.g., notion of brain-dead in humans); political (e.g., human abortion, nuclear energy); and other value laden factors. Situations such as these have given rise to bioethics as an academic field of study. While biological research has never been fully value free, it is less so today.

A notable change in biological research over the past several decades has been the emphasis on the study of humans in all aspects of their being and existence. For centuries, the biology taught in schools and colleges has been anti-human, a view fomented by research biologists who have assumed that the species was too complex to be studied scientifically. Up until now, it has been difficult for humans to be recognized as a species.

The development of modern biology with the direction of research toward a unified view of humans as a species has brought into question the traditional general biology or life science courses of schools and colleges. The charge is that these courses are "outdated," "obsolete," "outmoded," and "largely irrelevant" to human concerns, and that they should be completely reconceptualized. *America 2000* views this task of educational reform as one that will require a "quantum leap forward" with "far reaching changes" that are "bold, complex and long range."

There is little question that our nation is entering a new era—biological research is taking place in a new vein, and the adaptive demands for humans as life is now lived are changing. Each of these factors identifies purposes for the teaching of biology in schools, with a focus on helping young people to better understand themselves in a context that is as wide as life itself—biologically, cognitively, behaviorally, and sociologically. The educational task is one of developing a new synthesis of all that is known about humankind and how this knowledge can be interconnected in terms of human adaptive systems and real-life affairs.

Curriculum developers and science teachers of a modern biology program must recognize that the endeavor is one of invention with a clear vision of what it means to be a human. The revolutionary changes called for cannot be achieved by attempts to simply revise, restructure, reorganize, or update the current biology curricula; these courses and how they are taught are the source of the demands for educational reform. The task is one of divorcing ourselves from the past and developing new ways of thinking about a citizen's education in biology. As biologists and biology educators, our responsibility is to create a curriculum mutation that represents viable adaptive systems for humans in the natural and personal/social worlds.

2

Reforming Biology Education

Chapter 1 asserts that science education in the 1990s must be characterized by reform through practical action. This statement recognizes that
educational reform is more than writing reports and calling for action.
Rather, reform presents the opportunity to improve biology education by
acting on the numerous recommendations for improvement (Jones, 1989;
Rosen, 1989; Blystone, 1990; Kahle, 1990; Walker, 1990; Yager and Tweed,
1991; and Wright and Govindarajan, 1992). The most practical level of
action in the reform occurs in biology classrooms.

This chapter briefly reviews the problems of biology education, examines
different aspects of reform in high schools, two-year colleges, and four-year
colleges, and concludes with a discussion of the design and use of this guide.

What Are the Problems in Biology Education?

Recommendations for the reform of high school and college biology
education are not unique to the 1990s (Andrews, 1964; CUEBS, 1967), and
the history of biology education has witnessed several periods of substantial
reform (Hurd, 1961; Rosenthal and Bybee, 1987; 1988). Each period of
reform, however, has had distinctive problems that required unique solutions. The contemporary reform of biology education must include the
continuum of institutions from high schools (AAAS, 1989; Rosen, 1989;
NRC, 1990), to community colleges (NSF, 1989), small liberal arts colleges
(Carrier & Davis-Van Atta, 1987), and large colleges and universities (AAC,
1982; NRC, 1982; NSB, 1986; NSF, 1989; AAAS, 1990; Carter, 1990).

Fulfilling the Promise: Biology Education in the Nation's Schools (NRC,
1990) provides a thorough review of the problems facing pre-college
education in the life sciences. The list is familiar: low test scores, inadequate textbooks, limited laboratory-based instruction, and poor assessment. The NRC report leaves little doubt about the dismal condition of
biology education in high schools.

The National Science Foundation's report on two-year colleges (NSF, 1989) identified another major issue:

> The community, junior, and technical colleges are an important and integral part of American higher education....Over five million credit students, including 55% of first-time college students, are attending these two-year institutions. Also, 42% of black students, 54% of Hispanic students, 43% of Asian students, and over 50% of all women collegiate students are in two-year colleges.

Two-year colleges play an essential role in improving the biological literacy of all citizens and in providing recruits for the work force in professions related to science, technology, and health. Proposals for the reform of biology education, however, have all but overlooked the vital role of community colleges.

Problems in biology education extend to four-year colleges, as another NSF report (NSF, 1989) demonstrates:

> The main target for reform must be introductory biology in the colleges and universities, for this may be the terminal course for those who will become our national leaders, scientists in other fields, informed citizens, and those who aspire to a career of teaching in our nation's schools.

The foregoing problems point to a need to reconceptualize curriculum, instruction, and assessment for biology programs. For example, the National Research Council (NRC, 1990) proposed:

> ...the high school biology course should be a synthetic treatment of important concepts and of how these concepts can shape our understanding of ourselves and our planet.... We need a much leaner biology course that is constructed from a small number of general principles that can serve as a scaffolding on which students will be able to build further knowledge.... The scaffolding should include an understanding of basic concepts in cell and molecular biology, evolution, energy and metabolism, heredity, development and reproduction, and ecology. Concepts must be mastered through inquiry, not memorization of words.

Reform should produce a coordinated set of experiences for those passing through the respective educational levels, and the biology curriculum must accommodate the unique educational characteristics of students at each level. The curriculum should provide sufficient latitude for instructors to tailor the biology program to their unique needs and resources.

In addition, the ever-mounting problem of recruitment and retention of individuals for careers in science, engineering, and health (Vetter, 1989) requires special attention to the needs of women and individuals from traditionally underrepresented groups who may be interested in science. Underrepresentation is not just a matter of demographics, but a pedagogical problem (Johnson, 1984); pedagogical approaches that address cultural alienation to science need to be implemented for historically underrepresented people.

Taking a Closer Look: Why Reform Biology Education?

After a decade of reports, there is disagreement about what biology should be taught, how biology should be taught, and how to bring about changes in extant programs. Unlike multiple competing hypotheses in science, however, these disagreements have not served to promote progress. Instead, these different views result in diverse, uncoordinated attempts to improve biology education.

The curriculum framework proposed in this guide coordinates some of these diverse attempts by focusing on a more effective instructional approach for all students—eveloping biological literacy. This framework acknowledges and capitalizes on the views of those teaching biology in high schools, community colleges, and four-year colleges. The following discussions elaborate the general needs of science education and the specific needs of biology education at different educational levels.

High School Biology Education

The National Research Council (NRC) report *Fulfilling the Promise: Biology Education in the Nation's Schools* (1990) recommends the following for high school biology:

- In designing a course, we must identify the central concepts and principles that every high school student should know, and pare from the curriculum everything that does not explicate and illuminate these relatively few concepts.
- The concepts must be presented in such a manner that they are related to the world that students understand in a language that is familiar.
- Biology courses must be taught by a process that engages all the students in examining why they believe what they believe. That requires building slowly, with ample time for discussion with peers and with the teacher. In science, it also means observation and experimentation, not as an exercise in following recipes, but to confront the essence of the material.

These recommendations support a biology curriculum that uses unifying principles to present major concepts in a context that has meaning for the student. The NRC report also emphasizes the importance of instruction grounded in inquiry. The framework outlined in the following chapters accommodates these recommendations.

The omission of a conceptual framework in the NRC report is partially remedied in reports from the American Association for the Advancement of Science (AAAS). As part of *Project 2061*, AAAS outlined goals in science literacy that all Americans should achieve by age eighteen. The goals, described in *Science for All Americans* (AAAS, 1989), provide a view of biology content that contrasts markedly with that found in most current textbooks. A chapter on "the living environment," for example, focuses on six major subjects rather than a myriad of unrelated facts. Of particular interest is a companion volume, *Biological and Health Sciences: A Project 2061 Panel Report* (Clark, 1989). This AAAS volume outlines the conceptual themes, and suggests, but does not elaborate, a human context for the curriculum.

The National Research Council (NRC) is directing development of national standards for science education, a task that will make a significant contribution to the reform of science education. By fall, 1994, the project will publish Science Education Standards that represent the consensus of teachers and other science educators, scientists, and the general public about goals important for *all* students to attain in science education. A comprehensive set of standards will provide qualitative criteria to guide judgments about development of curriculum, teaching, and assessment of programs and students.

The standards, however, will not specify actual curricula, but will suggest design parameters for curricula, outline a variety of effective teaching strategies, and describe tasks, projects, and portfolios that teachers may use in their assessments of students. When published, the Science Education Standards should contribute to the coordination of, and provide a coherence for, the reform of science education, but will neither outline a national curriculum nor mandate federal guidelines for science education. The ultimate decisions on curriculum, teaching, and assessment will rest with supervisors, science educators, curriculum developers, and science teachers.

Undergraduate Biology Education: Two-year Colleges

Undergraduate biology education in community and junior colleges has been neglected. This unfortunate situation must not continue because two-year colleges are a vital extension of high school, a critical link to four-year colleges, and the institutions that provide terminal degrees for many students. The large number of students who attend two-year colleges (approximately three million), the underrepresented groups among those students, and the growth in numbers of two-year colleges (from 637 in 1964 to 1272 in 1984), make a particularly compelling case to include two-year colleges in any reform of undergraduate biology education.

Many students in two-year colleges take their last formal science course there, and that course is often biology. According to data available from the publishing industry (L. Loeppke, personal communication, August, 1990), approximately 280,000 students enroll annually in general biology courses at two-year institutions. Many students are pursuing occupational or technical programs and move directly from the two-year college to employment. Those who do transfer to four-year colleges usually complete their science requirements in two-year colleges (NSB, 1986).

A 1989 NSF Report on the National Science Foundation Workshop on Science, Engineering, and Mathematics Education in Two-year Colleges recommended support for curriculum revision and development of improved laboratories, including hands-on activities for introductory science courses. In addition, there was a recommendation for programs that enhance the use of computers and other educational technologies in introductory courses.

Undergraduate Biology Education: Four-year Colleges

One of the most comprehensive attempts to improve undergraduate biology education was initiated by the American Society of Zoologists as the

Science as a Way of Knowing project. The project has produced seven volumes that include updated knowledge in specific areas in biology such as evolution and human ecology. One goal of the project is to establish a conceptual framework for biology. John A. Moore, who conceived and directed the project, had this to say:

> Many of our best teachers agree that the essence of education should be to provide students with the conceptual framework of a field. Facts come and go—not necessarily because they are wrong, but because they become uninteresting or irrelevant. A conceptual framework, however, gives meaning to existing data and allows one to integrate new data and thoughts into a meaningful whole. The conceptual framework of a field is constantly adjusted and remodeled but it is never discarded. Once learned, it is not easily forgotten (Moore, 1984).

An NSF workshop on undergraduate biology education also recognized the need for a conceptual framework and recommended other areas that need attention: laboratories and course development, enhancing the quality of teaching, and recruitment and retention of students (NSF, 1989).

Conclusion

The BSCS designed and developed this guide for improving curricula in pre-college and college biology education in an attempt to resolve some of the aforementioned problems. We have synthesized the contributions of contemporary reports into a curriculum framework that is more specific than any report, yet this guide does not represent a fully developed program. The curriculum framework presented in this report is an intermediate step between the general recommendations of other reports and the specific details of programs developed by individuals or organizations. This guide includes background information and specific policies that local institutions or national groups can use as the basis for developing and implementing new biology programs. The framework specifically addresses the following challenges for biology education:

- improving the biological literacy of *all* students,
- increasing concentration on major unifying principles of biology and inquiry-oriented instruction,
- teaching biology in contexts that have personal, social, and ethical meaning,
- developing active learning environments through better use of discussions, laboratories, field experiences, and educational technologies, and
- providing implementation strategies for new programs.

This list constitutes the general goals for *Developing Biological Literacy*. The following chapters provide the details.

3

Understanding Biological Literacy

Educators use the phrase "scientific literacy" to express the major objective of contemporary science education, an aim recognized for *all* students. In this guide, the BSCS assumed the task of developing a curriculum framework for biology education in that context. We assume that translating the recommendations into an actual program for all students will contribute to the achievement of one aspect of scientific literacy, namely biological literacy.

Biology For All Students

Current school science programs do not provide opportunities for all students to continue learning either formally in colleges or universities, or informally as adults. For some students, often the underrepresented in science and technology and the underprivileged in society, program standards are low and the biology education inadequate. For other students, those at the opposite end of the spectrum, biology education is more difficult and advanced, but it is seldom directed to objectives that form a core of science education for all.

The contemporary phrase "science for all" implies an orientation for a general education that assumes individuals share significant relationships, in this case scientific, with others in a community. General education enhances and affirms the qualities of interdependence that unite us into one society. General education underscores science as part of our shared cultural heritage, a shared set of problems and possibilities related to science, and a shared future in which science will play an important role.

We use "science for all" to indicate that **any new biology program should address the needs and concerns of all students**, which, by definition, includes those who traditionally have been underrepresented in the sciences. The curriculum must provide a bridge between the intuitive, everyday experiences of different students and the counter-intuitive, grander-scale ideas of biology such as evolution. Although subject matter might be the same for all students, one must take into account the various prior experiences, understandings, and skills of those in the program.

The implication of "science for all" is clear: **biology programs should present the biological knowledge, skills, and values necessary for students to become effective citizens in society.** It is important to note, however, that science for all students does not mean that all classes should be the same nor that one class should be the same for all students. The following criteria for the design of a biology program follow logically from this implication. The content of the program should:

- be introduced with recognition of students' prior understandings, current achievement, and future potential,
- have immediate and evident personal and social meaning to the student and should enhance ethical decision making,
- extend and elaborate the students' understandings and skills,
- be taught in a variety of ways that demonstrate the processes, skills, and values of biology,
- be a challenge that is achievable at some basic level by all students and presented in a manner that promotes higher levels of biological literacy and open-ended learning, and
- be assessed in the context of personal, social, and ethical decision making.

A biology curriculum should set a bottom-line level of biological literacy for each student while striving for higher levels of literacy for all students. Science for all relates directly to the topic of biological literacy that we address next.

A Review of Biological Literacy: The Concept

Few biology educators have addressed the general topic of biological literacy. Ewing, Campbell, and Brown (1987) report a study in which college students read about bioethical issues from sources other than textbooks and engaged in discussions that made connections between those readings and citizenship. The research supported the investigators' hypothesis that such activities would encourage scientific literacy among students. The researchers recommended incorporating bioethical issues in the biology curriculum, a recommendation that unites many of the studies cited below. Jones (1989) used biological literacy—the public's lack of knowledge, understanding, and perspective about science—to argue for incorporating nature study into contemporary biology curricula. Gibbs and Lawson (1992) used the goal of scientific literacy as the basis for their study on the nature of scientific thinking as reflected by the work of biologists and the writing in high school and college biology textbooks. The authors concluded, not surprisingly, that the majority of Americans are not scientifically literate—if scientific literacy means understanding how scientists do science. Most textbooks do not portray an accurate view of the processes of science. Gibbs and Lawson suggest that the reason for this portrayal is that textbook authors are *themselves* not scientifically literate. In another report on biological literacy, Demastes and Wandersee (1992) suggested that the essence of biological literacy is understanding a small number of pervasive biological principles and applying them in appropriate ways to activities in personal and social spheres. The authors recommended modification in college biology courses to make biology more relevant to students.

Other authors have discussed various dimensions of reform in biology education but have not directly addressed biological literacy as a goal. Although biological literacy has not been thoroughly addressed by science educators, those who have dealt with this issue have argued that the nature of science and bioethical issues should be used in classes to promote biological literacy for all students.

Biological literacy requires an understanding of the nature of science. We consider it essential for students to know and understand something about the characteristics of scientific knowledge, the values of science, and the methods and processes of scientific inquiry. Table 3-1 identifies some of the important characteristics of a biologically literate individual.

A Biologically Literate Individual Should
Understand
<ul><li>Biological principles and major concepts of biology</li><li>The impact of humans on the biosphere</li><li>The processes of scientific inquiry</li><li>Historical development of biological concepts</li></ul>
Develop Appropriate Personal Values Regarding
<ul><li>Scientific investigations</li><li>Biodiversity and cultural diversity</li><li>The impact of biology and biotechnology on society</li><li>The importance of biology to the individual</li></ul>
Be Able To
<ul><li>Think creatively and formulate questions about nature</li><li>Reason logically and critically and evaluate information</li><li>Use technologies appropriately</li><li>Make personal and ethical decisions related to biological issues</li><li>Apply knowledge to solve real-world problems</li></ul>

Table 3-1: Characteristics of biologically literate individuals

Science as a Way of Knowing

The phrase "science as a way of knowing" concisely presents what we think students should understand about the nature of science. This phrase, used by John A. Moore in a series of volumes prepared for undergraduate biology programs (Moore, 1983; 1984; 1985; 1986; 1987; 1988; 1989;

1990), suggests that science is *a* way of knowing, not *the* way of knowing, and it conveys the idea that science is an ongoing process focused on generating and organizing knowledge.

Biological knowledge changes continually, and in recent decades, changes have occurred at increasing rates and along different paths. Biological knowledge is revised, expanded, and elaborated based on new observations, better instrumentation, or additional experimental evidence. Changes in biological knowledge are governed by the dynamic relationship among the existing body of scientific knowledge, the values of science, and the methods and processes of science. We briefly describe these influences in the next sections.

The Nature of Biological Knowledge

Most biologists and biology educators agree that a scientifically literate person should understand the nature of science. In this section, we have chosen five factors, among many, that illustrate the nature of biological knowledge and that should be incorporated into biology programs.

Biological knowledge is tentative. Biological knowledge is subject to change; the knowledge of biology is neither absolute nor complete. Science does not prove anything once and for all time; biological knowledge enables one to make probabilistic predictions and explanations for observations, organisms, and events in the natural world.

Biological knowledge is public. One assumption of science is that other individuals could arrive at the same explanation if they had access to similar evidence—the facts. There is, therefore, a requirement that biologists make public new knowledge and the methods used to discover that knowledge.

Biological knowledge is empirical. Observation and experimentation are the basis for scientific knowledge. The validation or refutation of scientific knowledge is based in the natural world without appeal to supernatural explanations. The test of knowledge claims is empirical validation.

Biological knowledge is replicable. Scientists working in different places at different times should be able to repeat another scientist's observations and experiments and derive the same evidence.

Biological knowledge is historic. Biological knowledge builds on and revises the accumulated corpus of understanding. Knowledge from the past is the basis for present scientific knowledge, which subsequently is the basis for future knowledge. Historical knowledge in biology must be examined in the context of the period, and not evaluated solely in light of contemporary understanding. Similarly, contemporary scientific knowledge should be understood in its social, technological, and political context.

Values of the Scientific Enterprise

Students should understand the principles, standards, or values that govern scientific work and that guide the acceptance of knowledge claims into the structure of biology. By implication, these values of science are considered to be worthy characteristics of a scientifically literate person.

Although understanding of biological concepts is one goal of biology education, another should be to develop an inclination in students to apply the values of science to the natural and technological world. The values of science that follow have been paraphrased from those listed in the report *Education and the Spirit of Science*, published by the Educational Policies Commission (1966).

Knowledge is valued. Knowledge and understanding are the primary outcomes of science. Scientists consider the investment of time and resources in the search for knowledge a worthy goal, and thus they place a high value on knowledge.

Questioning is essential. Development of scientific knowledge is based on questioning current knowledge. All knowledge that forms the conceptual schemes of biology is open to question. Authoritative statements by other scientists and beliefs of nonscientists need not be accepted as knowledge by any biologist. Scientists hold the position that all knowledge claims can be questioned; based on new knowledge, the extant concepts are subject to revision. In short, questioning leads to inquiries and subsequently to the revision of current knowledge.

Data are fundamental to biology. Statements of empirical fact, not arguments of belief or appeals to authority, resolve scientific disputes. The basis for scientific knowledge claims is data. Acquisition and ordering of data are fundamental to the development of scientific arguments and the construction of theories.

Verification is required. The validity and accuracy of scientific findings from observations, experiments, and theoretical claims are open to review. Scientists honor requests for similar observations by others, replications of experiments, and presentation of crucial proofs and claims in the case of theoretical explanations.

Logic is respected. Scientists respect the patterns of reasoning that connect empirical findings and construct scientific knowledge. Claims about new findings must be presented with logic and clarity.

Processes of Scientific Inquiry

Scientists use a variety of methods and processes in their work. In this section, we present five categories representing a synthesis of scientific processes.

Observing, classifying, and inferring. Observation, or the use of senses to obtain information about the world, is the basic process of scientific inquiry. There are numerous objects, organisms, and events that one can observe, so there is a need to impose order and organization on objects, organisms, and events. This is the process of classifying. Inferring is the construction of tentative explanations and cause-effect relationships based on observations and systems of classification.

Formulating hypotheses. Forming a hypothesis requires stating tentative explanations for causal relationships among objects, organisms, or events. The hypothesis is subject to expanded empirical testing and subsequent validation.

Interpreting data. Investigators attempt to find patterns and provide meaning in a group of data. Interpretation of data leads to development of additional hypotheses, the formulation of generalizations, or explanations of natural phenomena.

Designing experiments. This process includes planning the methods and procedures for gathering data to answer a question, develop an explanation, confirm a hypothesis, or challenge a theory.

Conducting investigations. In conducting an investigation, the methods and processes of science, the values of science, and the skills of the biologist in the actual development of scientific knowledge are synthesized. In addition, investigation includes other processes and skills such as communicating results, controlling variables, defining terms, formulating models, and making scientific findings public.

Biological Literacy

This report asserts that the pursuit of biological literacy is a continual, life-long process. Most discussions of biological literacy use the term as a goal that one either achieves or does not, that is, a person either is biologically literate, or is not. It is much more appropriate to recognize that each individual occupies a position somewhere along a continuum of biological literacy for different biological concepts. Accordingly, the task for biology educators is to move students to a different position along the continuum—a position that implies a richer understanding of biology. The following discussion describes a four-level continuum of biological literacy, clarifies some of the characteristics of students at each level, and suggests teaching strategies that will promote continued development of biological literacy. Table 3-2 characterizes of each of the four levels of literacy.

Nominal Biological Literacy	
Students:	identify terms and concepts as biological in nature, possess misconceptions, and provide naive explanations of biological concepts
Functional Biological Literacy	
Students:	use biological vocabulary, define terms correctly, but memorize responses
Structural Biological Literacy	
Students:	understand the conceptual schemes of biology, understand procedural knowledge and skills, and can explain biological concepts in their own words.
Multidimensional Biological Literacy	
Students:	understand the place of biology among other disciplines, know the history and nature of biology, and understand the interactions between biology and society.

Table 3-2: Characteristics of the four levels of biological literacy

Nominal Biological Literacy

Many students recognize the domain of biology and certain words as belonging to the realm of biology, as opposed to other disciplines such as art or political science. Students often come to class with a nominal level biological literacy, that is, they are literate in "name only." Students may have heard a biological term or concept; they even may have studied biology before but they did not develop an understanding of the information presented to them. As a consequence, the term has no meaning or importance to the students. They have what contemporary learning theorists describe as misconceptions, or naive theories. Table 3-3 lists examples of misconceptions expressed on the first day of an introductory college biology class in which students were asked to define given terms only if they were certain of their meaning.

- **algae**—a type of fungus
- **anthropology**—the study of plants
- **capillary**—something in your blood
- **carboniferous**—a total meat eater
- **carrying capacity**—how much you can carry
- **chaparral**—an animal
- **eocene**—some plants use this as food intake
- **epiglottis**—little pigments inside the plant
- **era**—the aroma or surrounding
- **finite**—forever
- **genus**—the reproductive part of a plant
- **germinate**—spraying to eliminate harmful air
- **locus**—an insect, or kind of plant
- **sampling**—small young tree
- **testosterone**—a male sex organ
- **virus**—a sickness caused by bacteria

Table 3-3: Selected misconceptions of college students on the first day of an introductory biology class

Although some humor can be found in the definitions listed in Table 3-3, it is also the case that all of the students' definitions are basically "scientific" in the sense that these students recognized the terms as scientific and tried to define them as such.

Functional Biological Literacy

Students may be introduced to a biological term or concept through their reading or by an instructor or a television program telling them about it. This introduction may serve as the only contact the students have with the subject. Students may memorize an appropriate definition of the biological term or concept, which leads to *functional biological literacy.* At this level, students are able to define certain biological terms or concepts but have limited understanding of or personal experience with them. Biology programs that emphasize vocabulary and rote memorization lead to a functional level of biological literacy. Unfortunately, this is often the emphasis of biology programs and the point at which textbooks and instructors stop, leaving students with only one dimension of biological literacy. Students at this level have no understanding of the discipline, no feeling for the excitement of scientific investigations, and no interest in biology. Simply reading about biology may help inform students about some natural phenomenon, but only at a functional level. Just as people described as functionally literate can "get by" in life, students who have a functional biological literacy may be able to get by on certain objective examinations. Teaching and learning at the functional level recognizes and supports the importance of biological vocabulary, but educators should help students develop higher levels of biological literacy.

Structural Biological Literacy

If students are to develop a higher level of biological literacy, they must understand the major conceptual schemes of biology, those ideas that help organize all of biological thinking. Table 3-4 lists the unifying principles of biology, which are elaborated in Chapter 5. These principles provide a conceptual framework for organizing biological content.

Evolution: Patterns and Products of Change
Living systems change through time.
Interaction and Interdependence
Living systems interact with their environment and are interdependent with other systems.
Genetic Continuity and Reproduction
Living systems are related to other generations by genetic material passed on through reproduction.
Growth, Development, and Differentiation
Living systems grow, develop, and differentiate during their lifetimes based on a genetic plan that is influenced by the environment.
Energy, Matter, and Organization
Living systems are complex and highly organized, and they require matter and energy to maintain this organization.
Maintenance of a Dynamic Equilibrium
Living systems maintain a relatively stable internal environment through their regulatory mechanisms and behavior.

Table 3-4: Unifying principles of biology

One can think of the unifying principles as branches on a tree that has evolution as its trunk and the "facts" of biology as its leaves. An understanding of the trunk and branches of the biological tree constitutes structural biological literacy, whereas functional literacy focuses on the leaves alone—and when the leaves fall, the facts are forgotten. Continuing the tree metaphor, for students to reach a structural level of literacy, they also should understand the processes of growth for the tree, that is, the nature and methods of scientific inquiry.

We assume that students are willing to learn more about biology if it is meaningful and interesting to them. At least some students must overcome a barrier of disinterest before they study biology. One way to engage students is to use hands-on investigative experiences that can generate excitement and enthusiasm while introducing students to biological concepts. In addition, teachers should help students see how biological concepts relate to them. For instance, although photosynthesis may be taught as a series of chemical reactions represented by a chemical equation, students should learn and appreciate the significance of the process to their lives—that the oxygen they breathe and the food they eat are byproducts of photosynthesis. Similarly, memorizing the phases of mitosis provides no understanding of the significance of cell division; relating uncontrolled mitosis to the development of cancer, however, can interest students in mitosis because the study of cancer is more interesting and relevant to them. Identifying the personal significance of biological concepts addresses the common student question, "Why do I have to study this?" All concepts in biology either have a *direct or indirect* relevance to the lives of students, and highlighting these relationships by using current events, interesting natural phenomena, or human-related examples should promote an increased appreciation for their study.

Once students are engaged, they may seek, or at least assimilate to a greater extent, concepts and skills that will help them better understand a particular subject. Students may begin to construct their own meaning of a subject based on sources such as readings, laboratories, and research. Here, directed-inquiry laboratories can help students discover information for themselves while they learn the processes of scientific investigation. Empowering students with tools to pursue investigations of natural phenomena on their own is one way to promote biologically literacy. The following paragraphs provide students with one such set of investigative tools.

Asking the Right Questions. To investigate any natural phenomenon or problem, students need to ask the right questions. First-year journalism students learn to ask "Who, What, Where, When, Why, and How?" to investigate news stories. Similarly, the processes of science are not mystical processes that only a few people can undertake; there are a few, general questions that one can ask to learn more about any aspect of biology. Mastery of these questions can empower students to pursue investigations on their own and to develop higher levels of biological literacy throughout their lives.

By asking a combination of questions listed in Table 3-5, students may begin a study on their own. The first group of questions should be asked when a natural phenomenon or problem is first encountered. The second group of questions will help students explore the phenomenon or problem, and the third should help students think about the future. The last group of questions allows students to assess the validity of information used to explain the phenomenon or problem. By using these questions, anyone can conduct investigations in a scientific manner.

Description
• What is it?
• What does it do, or what is its function?
• Is it typical?
• Is it real or an artifact?
• To what is it related?
• What is its significance/importance/advantage?
Investigation
• Can it be investigated scientifically?
• How can I find out about it?
• How does it work/happen?
• How does/did it begin?
• What causes it to happen? (Is there a cause/effect relationship, or just a correlation?)
• What affects it, and what does it affect? (How does it fit into the whole picture?)
Prediction
• What are the expected results?
• What would happen if _____?
• Has it changed or does it change through time?
• What are its future risks, benefits, or consequences?
• Does it always happen like this?
• Can knowledge about it be used to solve problems or be applied to other situations?
Evidence
• What is the evidence?
• Is the evidence based on false assumptions?
• Are the data/facts about it accurate?
• How do I support my position about it?
• What other questions do I need to ask about it?
• Are there any other explanations?

Table 3-5: Inquiry questions for investigating biological phenomena

A student reaches *structural biological literacy* when he or she understands a subject well enough to explain it to another person **in his or her own words**. At this level of literacy, a student understands how the information was obtained, is able to apply information about the subject to novel situations, and knows and appreciates the significance of the information to biology and to himself or herself. The structural level of biological literacy is a foundation on which the understanding of other, related biological concepts is based. For instance, a student may be sexually active and develop an interest in sexually transmitted diseases. He or she may develop an understanding of the cause of AIDS and thus be able to understand why certain behaviors are more risky than others, why the virus cannot be controlled with antibiotics, and why a cure for the disease is not on the horizon. This understanding, in turn, may lead to learning about other health issues, about differences between bacteria and viruses, and about biomedical technology.

Multidimensional Biological Literacy

This level of biological literacy represents a broad, detailed, and interconnected understanding of a subject area of biology. Each area of biology, for example, has a history, is influenced by a variety of social and global issues, may incorporate technology, and is related to conceptual schemes. As students reach a fuller understanding of a biological subject and its connections to other subjects and disciplines, they develop multidimensional biological literacy. Few professional biologists, not to mention students, achieve this level of literacy in *all* areas of the discipline, hence the view that biological literacy is a continuum. A person may have multidimensional literacy about the life histories of temperate birds, but have only a functional level of literacy about the molecular biology of plants.

Students may reach the multidimensional level if they continue their study of a subject, if their personal interest remains high, or if they are interested in a subject and are confronted with a problem related to it. For instance, a student may be functionally literate about global warming. If, however, he or she faces the possibility of giving up driving because automobile exhaust contributes to global warming, he or she may see a need to act on the problem, or at least to discover more information that will help make informed decisions about whether to drive or not. This may be followed by a willingness to act or to investigate further the issue of global warming.

Students may not be willing to organize and undertake a plan of action to investigate a problem. Such students may be drawn into action if they develop a commitment or a relationship to the subject. For instance, in northern California, students who studied a bog as a class assignment developed a fondness for the natural area, and when a developer proposed destroying part of the bog to construct a parking lot, the students became committed to saving the bog. They were willing to protect the bog and began a plan of action *on their own* that required detailed information about the bog's physical and biological attributes as well as its importance to members of the human community.

Multidimensional literacy involves the ability to investigate a problem concerning a local issue or a biological concept, to collect information related to the problem or concept, and to apply this knowledge to the solution of the problem or to the expansion of biological knowledge. Students discover that this intensive investigation requires the interconnection of many ideas and much information. Once they begin an intensive, long-term study of a subject, students should recognize the types of information and skills required to answer a question or solve a problem, and they should realize that biological knowledge must be integrated with knowledge from other scientific, mathematical, social, political, and economic disciplines. The multidimensional level of literacy cultivates and reinforces life-long learning in which individuals develop and retain a need to know and have acquired the skills to ask and answer appropriate questions. Figure 3-1 illustrates the biological literacy model described in the previous pages.

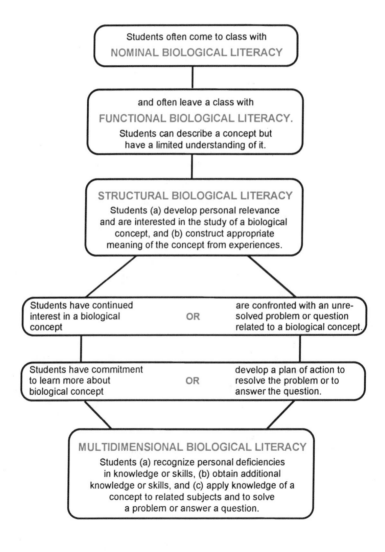

Figure 3-1: Biological literacy model

Conclusion

Biological literacy has several dimensions such as understanding concepts, using vocabulary, and developing a larger view of the biological sciences. We see biological literacy as a continuum along which individuals continually progress. We also envision structural and multidimensional levels of biological literacy as goals for all students. This chapter concludes with an essay by John A. Moore titled "What is Natural Science?" The essay presents an excellent overview of what the biologically literate citizen should understand about the natural sciences. Dr. Moore's understanding of natural science clearly is multidimensional.

ESSAY

What is Natural Science?

by John A. Moore

The word "science" is derived from the Latin *scientia*, meaning "knowledge," and that broad meaning has been used since the Middle Ages. Thus, in addition to the natural sciences we have culinary science, military science, theology (once regarded as the "Queen of the Sciences"), social science, and Christian science. The *Oxford English Dictionary* gives that broad usage as the first meaning. One must go to section 4a in that dictionary to encounter the more restricted usage, "A branch of study which is concerned either with a connected body of demonstrated truths or with observed facts systematically classified and more or less colligated by being brought under general laws, and which includes trustworthy methods for discovery of new truths within its own domain." In the late nineteenth century an even more restricted usage became common, namely, science is the study of living and nonliving nature and the body of information resulting from that study—the natural sciences.

The natural sciences included the study of celestial bodies, the fine structure and interactions of matter, the lands and waters of the Earth's crust, and all creatures living today or in the past—that is, except for some aspects of human life. Humans are, of course, living creatures, but they have added to an anatomy and physiology a nervous system of unique complexity. Thus, the field of animal behavior, as applied to humans, becomes economics, psychology, sociology, philosophy, technology, religion, and a host of other disciplines. When a bird sings, it is animal behavior; when (some) humans sing, it is music. Likewise, when a beaver builds a dam, it is animal behavior, but when a human builds, it is technology. About the only aspect of humans that we leave to biology is human anatomy and physiology and even these are sometimes referred to the domain of the medical sciences.

So the natural sciences here are restricted to astronomy, physics, chemistry, geology, and biology. Each of these separate sciences consists of an integrated body of information and of methods for acquiring information. There are unique methods for acquiring new information in each of the natural sciences but there are, as well, general procedures that apply to all. First there is the desire to understand some natural phenomenon, and this usually takes the form of a question. What is the explanation for those layers of rocks in a cliff? What is the function of that organ in a chicken?

Careful observation is the first step in seeking answers. Examination of the sedimentary layers might reveal different types of fossil organisms and different types of rocks. The organ of the chicken might be found to contain food material or be connected with the rest of the organism by blood vessels, nerves, and possibly by a tube to the bladder.

In both cases a second step would be to relate what was discovered about the sedimentary rocks or the chicken's internal structure with the body of available information. The facts of science have little relevance in themselves. Facts and observations only become important when related to the existing corpus of science. The sedimentary rocks could be related to the geological column consisting of all sedimentary rocks, and the contained fossils could be related to those found in other layers and in other places. The composition of the rocks might give a clue to how the layers were formed and, possibly, something about the climate at the time. The organ of the chicken would be better understood when seen as part of the total structure and physiology of the bird.

Routine observations have their limitations. More might be learned if a natural object were to be studied closely at higher magnifications. Or even lower magnifications—valuable information about the gross topography of a region can be provided by photographs taken by a satellite. Much of the body of scientific information has been acquired by observation alone. Systematic biology, for example, is based largely on descriptions—observations. New species of animals and plants are collected and their sizes, shapes, morphology, and, where possible, habitat and behavior are described. Again this is not an isolated activity. The

new species are classified, that is, related to the species previously described. Thus, over time, the catalog of nature becomes more complete.

Observation alone may provide only limited information and generally answers only the question, "What is it?" If one wishes to answer the question, "What does it do?" other procedures must be employed. The most powerful of these is experimentation. Experimentation consists of identifying all known factors influencing a phenomenon and then varying one while keeping the others constant. Thus, if a disease is suspected to be caused by a vitamin deficiency in a diet, diseased individuals are assigned at random to two groups. The control group continues with the standard diet while the experimental group receives increased amounts of the vitamin. The person conducting the experiment then observes and records the results.

So observation and experimentation are the initial methods of discovering the nature of a scientific phenomenon. It must be emphasized how difficult it may be to know what observations to make or experiments to perform. Time and time again a seemingly direct attempt to answer a question proves totally unprofitable and progress depends on some chance observations seemingly unrelated to the analysis. Thus, the underlying causes of biological diversity remained unknown for millennia until Charles Darwin speculated about the causes of geographic variation in birds of the Galápagos Islands and of the relation of fossil and living sloths in Argentina. The physical basis of inheritance remained an unanswered question until Walter S. Sutton pointed out that the chromosomal events in mitosis, meiosis, and fertilization paralleled the behavior of Mendel's factors in crosses of peas. It could be concluded that inheritance is based on chromosomes. The discovery of the planet Pluto was not made by first searching the heavens but by realizing that the orbits of known planets were slightly distorted. The cause of the distortion was suspected to be an unknown planet, its approximate position was predicted, and then the telescopes were trained on the predicted place—and there was Pluto.

Because progress in understanding natural phenomena depends so much on chance discovery, scientists must make large numbers of observations and perform many experiments in the hope of finding a clue that will allow progress. Thus the data base of science increases enormously while major conceptual advances come slowly.

The observations that can be made and the experiments that can be performed are very much a product of the climate of the times and the culture. The precision of the scientific question is determined by the level of the development of the field as well as other questions being asked. At any one time, a community of scientists is generally asking the same sorts of questions in much the same way, and the quest for answers will follow similar paths. A consequence of this is simultaneous discovery. A classic example is that Charles Darwin and Alfred Russel Wallace proposed the same hypothesis at the same time to explain organic diversity.

The natural sciences have advanced so much in the last century that current theory can even suggest the possibility of unknown phenomena and, hence, point to new areas for observation and experimentation. Science as a way of knowing is autocatalytic.

The extraordinary advances in our understanding of natural phenomena made by the sciences during the last century may suggest that special procedures—scientific methods—have been employed. Philosophers of science, beginning with Sir Francis Bacon, have identified the key features that lead to scientific discovery. It is an intellectual game. In its ideal formulation, a question is first recognized. Next a tentative answer to the question, a hypothesis, is proposed. If the hypothesis is assumed to be true, certain consequences, or deductions, must also be true. The de-

ductions must be testable, and if they are confirmed, then the hypothesis is made more probable. If the deduction cannot be confirmed, then the hypothesis must be abandoned or modified to account for the new information.

This is not a final truth and, indeed, science makes no pretense of dealing with final truths. The statements of science reflect only the best answer that can be provided for a scientific question relative to the time, sophistication of the field, and methods and technology available for obtaining data.

In spite of this philosophical caution, many statements in science have been confirmed for all practical purposes. The sun does "rise," the stars and planets do move in predictable ways, the facts of human anatomy permit surgery, and known genes are inherited in predictable ways. In many situations the "iffiness" of science moves to the background.

Confirmable data then are added to what is already known, and one can recognize the total body of information, hypotheses, and procedures relating to a broad natural phenomenon as a theory. Theories, then, represent the best statements that can be made about a natural phenomenon at the moment, and those statements can be expected to change over time as the phenomenon becomes better understood. When "theory" is used in this way, it differs importantly from "hypothesis," which is a tentative answer to a question that is being tested. Not infrequently "hypothesis" and "theory" are used as synonyms, which dilutes the value of each.

The elegance of scientific methods, which consist of a continuum of question—hypothesis—deductions—test of deductions—question—hypothesis—and so on, may lull one into assuming that they make up the recipe for continuous progress. Not at all. Questions that seek explanations for a phenomenon must be precise and practical. In the early days of physiology, the question, "How does the vertebrate kidney work?" would have been essentially useless. It would have been necessary to ask many little questions that would have answered questions about the gross and microscopic structure of the kidney and blood vessels and nerves that serve it. This information could be gained by observation. The next level of questions would be aspects of "How does the kidney produce urine?"

Even that question would have to be broken down into smaller ones such as "What is the composition of blood compared with urine?" or "What does the nephron do?" or even more specific, "What does the glomerulus do?" These questions all involve experimentation and are dependent on deep knowledge of chemistry and biology.

Scientists accepted that some hypotheses simply cannot be tested at a specific time because the techniques are not available. If the questions are important, however, a dogged attempt to develop the necessary techniques generally ensues. And the element of luck, combined with broad knowledge, may permit the scientist to continue the analysis. The chance discovery of the unusual structure of the nephron of the goosefish made it possible to answer questions about the function of the glomerulus in producing urine. The discovery of the giant axon in the squid permitted far more sophisticated experiments to discover the nature of the nerve impulse.

Thus the quest for deeper understanding of a natural phenomenon involves recognizing the phenomenon, asking an answerable question about it, proposing a tentative answer, deducing the consequences of what could be expected if the hypothetical answer is true, observing and experimenting to see whether the predicted consequences occur, reaching a tentative conclusion, having that conclusion confirmed, modified, or rejected by others, and making the conclusions available to all other scientists. At every stage, what is accomplished will depend on existing knowledge of the phenomenon and the tools and methods available for the

analysis. The infrastructure for scientific discovery requires an educational system and a society willing to support the expenses of scientific research.

And, of course, there is the human element. The scientist must attempt to be a detached observer and experimenter. That is, the task is to observe accurately and report honestly. Every effort must be made to avoid seeking an answer that one hopes to be true, such as confirming a personal belief. But no scientist can fully exclude personal beliefs and the seemingly established conclusions reached by others. This is why all important statements must be confirmed by others who, scientists hope, will not have the same constellation of unknown biases and other inadequacies.

Every step in the idealized sequence of major scientific methods is difficult and uncertain, so it is not surprising that progress, in answering major scientific questions, is usually intermittent. With such uncertainty of knowing what to do and how to do it, it is not surprising that progress is usually made by those who think more clearly, are broadly knowledgeable in the sciences and cognate fields, work the hardest, and have the infrastructure that makes research possible. Although genius does hold center place, an important discovery may be the result of chance, as many have been, but as Louis Pasteur said, "Chance favors the prepared mind."

Although the difficulties in developing answers may be great, a tenet of science is that all the phenomena of nature are, in principle, knowable. That means that all phenomena are thought to be based on things and causes that obey naturalistic laws of nature. It means also that supernatural forces cannot be invoked as part of explanations. Rain may be the tears of gods but that assumes that there are gods. Such a hypothesis, however, is untestable. That does not mean it is wrong, but any hypothesis that cannot be tested is useless. A scientist can have nothing to say about gods, or other assumed supernatural forces, and he or she must seek naturalistic explanations for the causes of rain.

Thus, slowly and painfully, new information becomes available that allows us to understand natural phenomena better. Understanding consists, for the most part, of relating one natural phenomenon to another or expressing it in mathematical terms. The vertebrate kidney is understood by relating what is known of its structure with the physiology of urine formation. Knowledge of the kidney of one vertebrate species can be compared with what is known of the kidneys of other species of vertebrates. Deeper knowledge of the phenomenon of excretion is obtained by comparing vertebrate kidneys with the excretory organs of species of other phyla.

Thus, the individual statements of science become parts of theories that provide explanations for the phenomena of science. And the most important aspect of these statements is that they must deal only with nature. The kidney is understood in relation to its structure and function and the structure and function of the entire organism. Science must interpret nature in nature's terms and supernatural explanations cannot become a *deus ex machina* to "solve" a difficult scientific problem.

This brings us to a comparison of science as a way of knowing with another dominant way of knowing—that provided by religion. The conflict with these two modes of explanation has been especially apparent in recent years with the attacks of fundamentalist Christian sects on evolutionary biology. They have proposed creationism as an explanation for the origin and diversity of life and its history over time. The fundamental difference is the sorts of evidence employed to explain the origin and diversity of living creatures. The basic evidence for a creationist is the book of Genesis in the Judeo-Christian Bible. One account of creation, the so-called Priestly version, has a god creating the world and all its inhabitants in six days. That evidence has to be accepted, if it is, on faith alone. There is no way of confirming whether or not there is a god or if that is what he or she did. A scientist cannot accept unprovable statements and so must seek other ways of explaining the data. In the last century satisfying naturalistic explanations have been provided by the fields of geology, evolutionary biology, genetics, and molecular biology. The gods may exist, but it has not been necessary to evoke them to account for the data. The phenomena of nature seem to proceed according to basic laws that the gods cannot, or do not, annul. That being the case, science is possible.

4

Thinking About A Contemporary Biology Program

The design, development, and implementation of a new biology program involves more than selecting and organizing subject matter. Although biological concepts are essential parts of a biology program, one must consider other factors such as the role of the laboratory, the use of computers, the assumptions about students as learners, and the use of materials such as textbooks in the design of a biology program.

This chapter presents information on topics we consider critical for those who are designing and implementing biology education programs. We use the term "thinking" in the chapter title to suggest that this information provides a framework from which one could design a biology program. Chapter 6 addresses the practical aspects of changing the biology program.

What Biological Education Should Achieve

The aim of biological literacy must be translated into specific goals for biology education. Education in biology should sustain students' interest in the natural world, help students explore new areas of interest, improve their explanations of biological concepts, help them develop an understanding and use of inquiry and technology, and contribute to their making informed personal and social decisions. Students should learn and understand how to use biological information in their daily lives. More specific goals are embodied in answers to the question: "What should biology education achieve?" The following paragraphs outline five goals for biology and science education.

Goals for a Biology Program

(1) Biology educators have continually based programs on a goal to develop fundamental scientific, technologic, and health knowledge (Rosenthal & Bybee, 1989). Developing Biological Literacy emphasizes the development of fundamental biological knowledge organized by major unifying principles in biology.

(2) Biology education should **develop an understanding of, and ability to use, the processes and methods of scientific inquiry**. Although descriptions of this goal have changed during the last 200 years, the goal has remained essentially the same. A review of literature about this goal shows that some investigators mean *understanding* scientific methods, some mean *using* scientific methods, and others mean skillful manipulation of equipment and use of technology (Bybee and DeBoer, 1993). Again, virtually all programs, syllabi, and materials in biology education include this goal.

(3) Biology education should **prepare citizens to make responsible decisions about social issues that relate to science, technology, the environment, and health**. Because science and technology are integral to our society, a contemporary biology curriculum should include a social component.

(4) Biology education should attempt to **meet students' personal needs**. National and state-level documents reviewed in preparation of this guide emphasize the goal of personal development, an important goal for all levels, but one that is especially important for high school students who may not continue to college. Aspects of science education such as critical thinking and reasoning, problem solving, scientific attitudes, and values also lend themselves to the personal development of students at the college level.

(5) Biology education should **inform students about careers in science**. Biological research and technological development continue through the direct work of scientists and engineers and through the indirect support of society. Knowledge of the variety of careers available to students has been a long-standing goal of science education, especially at the high school level.

Biology for All: What Is Common?

Scientists investigate the natural world. In any investigation, a scientist must make many decisions, including the determination of the phenomenon to be investigated, the methodology, the instrumentation, the protocol, the collection and reduction of data, the assessment of the uncertainty of the results, the correlation of results with existing knowledge, and the analysis of the theoretical implications of the results (Sigma Xi, 1989).

Inquiry

Students also may use processes of investigation in the classroom to gain an understanding of biological phenomena, concepts, and relationships.

Various educational task forces have stressed the use of scientific inquiry and reasoning in instruction (NCEE, 1983). Inquiry is a pedagogical method that combines laboratory activities with student-centered discussion and discovery of concepts. Students construct an understanding of concepts through the interaction of students with educational materials and the instructor. The teacher acts as a catalyst for learning, directing student interaction, activities, and discussion rather than simply presenting information to students (Uno, 1990).

Inquiry does not necessarily lead students to scientific discoveries, but the process does encourage students to discover information new to *them*. Inquiry is based on methods of scientific investigation that may include some or all of the following components: observing, questioning, forming hypotheses, predicting, experimenting, analyzing data, and relating ideas and concepts to each other.

Inquiry represents a radical alternative to conventional texts and lecture classes that are composed mainly or wholly of a series of unqualified, positive statements giving students the impression that science consists of a complete set of unalterable, fixed truths. In reality, science is continually restructured as new data are related to old. Inquiry also includes a fair treatment of the incompleteness of science and emphasizes the likelihood that with further inquiry, scientific knowledge will change (BSCS, 1963; 1978). The essence of using inquiry is to have students investigate questions and learn how to generate, analyze, and use data.

Biology for All: What Is Unique?

High school, community college, and four-year college nonmajors biology courses have unique student populations. Because the design of a biology program should be based, in part, on an understanding of the students in each type of course and their unique needs, it is important to recognize the differences among populations of students.

High School Students

High school is a time to reach *all* students. Of high school, community college, and four-year college/university students, the high school population is the most diverse. The high school student population has a high proportion of minority students—although such "minorities" will make up the majority of U.S. citizens by 2000 (BSCS, 1988)—and students from diverse racial, ethnic, and socioeconomic backgrounds. In addition, these students collectively possess diverse learning abilities and preparations for studying biology. The high school student population contains some unmotivated students who lack an adequate science background as well as a high percentage of potential dropouts. Approximately half of all high school students will take no further science courses after tenth-grade biology (Hurd, et al., 1980; CCSSO, 1990).

The objective of science education programs in the 1990s and beyond is to prepare future citizens for their role in a world shaped in part by an explosion of knowledge in biology and biotechnology. Citizens must be able to analyze scientific information from the media and make informed, ethical decisions about social issues that affect them and society at large. High school biology courses, therefore, must be structured to give students practice in the skills they will need to function effectively as citizens in the twenty-first century. Specifically, students need to *explore* science while developing all aspects of biological literacy. Accordingly, by graduation, high school students should be able to:

- demonstrate an adequate knowledge of major biological concepts and processes,
- access a variety of biological information for practical purposes such as interpreting directions on medicine bottles and nutrition information on prepared food packages,
- demonstrate the application of biological principles in their lives,
- value the diversity of life and its relationship to human endeavors,
- recognize and accept individual responsibility for local human activities and their impact on the global environment,
- think critically, for example, recognize the difference between opinion and fact,
- use appropriate procedural skills for scientific inquiry and problem solving, and
- solve problems cooperatively and creatively.

Community College Students

The student population in community colleges is unique because of the combined effect of a large percentage of minority students, adults who are returning to complete a college education, immigrant adults who use education to obtain jobs and to assimilate into American society, and those who will enter the work force immediately after completing community college. For many students in community college, their college degree represents their last formal education; a small number of these students continue their education at four-year colleges and universities. Like the high school population, this student population includes some who are unmotivated and who lack basic skills and an adequate science background. The unique constituency of students in community colleges has goals that include obtaining an education that enhances their lives or professions, training or retraining for a job or career, and preparing for further advanced studies. Seminars in study skills, basic reading and writing skills, overcoming science anxiety, and career awareness are especially appropriate for community college students. Community college biology students should be able to:

- find relevant information and recognize biases and errors of fact in information,
- make personal and social decisions concerning biology,
- critically evaluate issues of human health, environment, and societal implications of biology,
- understand and use scientific methods of investigation, and
- demonstrate basic knowledge about the life sciences.

Nonmajors in Four-year Colleges and Universities

The population of four-year nonmajors consists of students whose goals do not include a career in biology or related fields. They often are unmotivated by science in general and biology in particular. There usually is a narrower and higher range of ability among these students than among students in high school and community college courses. Four-year college nonmajors may be interested in learning but not necessarily in learning about biology. In addition, because faculty members often are teaching hundreds of students at one time, individual or personal interactions between faculty and students are rare, resulting in the alienation of some students. Nonmajor students have few opportunities to make connections between classroom information and the world outside and to participate in scientific investigations (Project Kaleidoscope, 1991). By the end of the introductory course in biology, nonmajors should be able to:

- appreciate and value science,
- make informed decisions about science-related personal and social issues,
- read, understand, critique, and discuss popular scientific information,
- justify and defend positions on science-related issues,
- apply scientific processes of investigation to daily life,
- make sense of the natural world,
- understand the major principles of biology and science as a process, and
- understand the role of humans in the biosphere.

Biology for All: What Are the Components?

The understanding of biology a student takes from any biology program depends on the content to which he or she is exposed; personal interest, motivation, and capacity to learn; and the instructional strategies used in the classroom. Four major strategies of biology programs are laboratory and field activities, educational technologies, textbooks, and lectures. Each of these is discussed below.

Laboratory and Field Experiences

Laboratory experiences are important educational experiences for *all* students because they accommodate individual differences in learners. Laboratory activities allow for exploration and discovery, essential parts of learning for any student, but need not be confined to buildings—field experiences in natural areas and trips to zoos and botanical gardens can be adapted into investigative experiences. The laboratory experience is so integral to the nature of science that it must be included in every science program for every student (NSTA, 1991). The laboratory experience should help students develop their investigative, organizational, creative, and communication skills. NSTA notes the critical importance of laboratory activities that enhance student performance in such skills as data analysis, communication, and conceptualization of scientific phenomena.

All biology programs should incorporate a variety of instructional strategies, and laboratory and field experiences should be central components of every program. Such experiences are opportunities for students to discover major concepts and to experience and use the processes of science. To accomplish these goals, laboratory and field experiences should include investigations that are exploratory and open-ended and that originate with student questions. These experiences should allow students to make observations, develop hypotheses, design and conduct experiments, collect and analyze data, and communicate their proposed answers and explanations. These experiences should go beyond confirming what students already know or requiring students to follow well-defined directions. Laboratory investigations also are perfect opportunities to demonstrate how the unifying principles help to organize biological content, as, for instance, in a comparative anatomy laboratory, where students discover patterns of evolution in homologous body parts of diverse animals.

Students should be challenged to answer questions based on authentic biological investigations and encouraged to ask, "What is the evidence?" When used as the basis for investigations, case studies and local public issues allow students opportunities to gather real data for analysis, to be challenged by both the qualitative and quantitative aspects of investigation, and to focus less on getting a "right" answer and more on the process itself.

Educational Technologies

Interactive educational technologies, like laboratory experiences, accommodate individual differences in learners (Hawkins and Sheingold, 1986). A major recommendation of this guide is that **all biology programs should incorporate educational technologies in ways that are appropriate to biological inquiry, adequate to program requirements, and that accommodate diverse learning styles.** Educational technologies can help students acquire information, use knowledge to make informed decisions, and access the vast amount of scientific information available in primary and popular literature (Papert, 1981; Pea, 1984; Linn, 1987). Videodiscs and videotapes can replace pages of print material, and the use of such visual media can engage most students because the mass media have a positive influence in students' lives and because many students are visually oriented. Another component of modern educational technology is the computer, which is a patient, nonjudgmental, and persistent tool for interactive learning and which provides students with instant feedback to their responses. In a laboratory setting, the computer can gather data independently and manipulate the data qualitatively by sorting and storing the information and quantitatively by computing descriptive and inferential statistics. The computer also is an effective tool for modeling complex systems, thus extending students' ability to understand interactions within and between systems.

The computer is a powerful tool for interactive learning because it can be programmed to prompt the student to enter specific data or responses needed for a particular learning activity and because it requires the student to process the information presented to him or her. Further, when the computer program allows the student to control the pace of instruction, frustration is reduced and opportunities for successful learning are increased. By evaluating the student's input and providing a useful response, the computer provides feedback that encourages the student to engage in the learning process and gives clues for modifying his or her thinking. Below we list some specific educational technologies and their appropriate uses.

Interactive videodiscs provide a visual database that allows students to explore relationships among living and nonliving things, to ask "what if" questions, and to learn science by responding to directed questions. Interactive videodiscs also can present simulations and visual representations of phenomena and tutor the students in solving problems and in developing scientific concepts and skills. Interactive videodiscs are useful for: rapidly illustrating many different organisms, structures, or processes; comparing and contrasting visual images; presenting organized data in the form of graphs, charts, and tables; providing background and historical information about biological concepts; allowing students to control and access information for their own research or presentations; and allowing teachers easy access to information for a variety of subjects.

Microcomputer-based labs (MBLs) enable students to obtain and record data with a high degree of accuracy over short or extended periods of time. Various sensors collect data on environmental conditions such as pH, light, and temperature. These sensors are connected to a "black box," or interface, which in turn is connected to a microcomputer. Data collected by the sensors are fed through the interface into the microcomputer, which displays the data in digital, graphic, or print format. A laboratory activity built on a specific interfacing system that includes hardware, software, data files, and procedural instructions constitutes a microcomputer-based lab. Microcomputer-based labs encourage students to think scientifically (Mokros and Tinker, 1986; Tinker, 1984). Because they see a graph in real times—the data are being recorded—students can analyze the data immediately, modify their hypothesis, or construct a new hypothesis and try another experiment. Without MBLs, data analysis usually occurs after the laboratory session, and there is less opportunity to test a new hypothesis. The ability to repeat or modify a procedure promotes the development of the concept being investigated. MBLs are useful for: providing alternative hands-on experiences for students; acquiring and displaying data in real time; allowing greater time for data analysis; presenting data in different forms; and allowing students to acquire data on a long-term basis.

Computer simulations are models of actual phenomena. An example of a simulation might be the physiological interactions involved in running a marathon. A simulation usually simplifies the system under study so that students focus on one or a few aspects, for example, the relationship of glucose use, glycogen depletion, and lactic acid buildup to the rate of running. Simulations provide students the opportunity to investigate problems that otherwise would be impossible to study because of cost, location, or facilities. They allow students to develop models of natural phenomena, to check their models against reality, and to manipulate models of natural systems in ways not possible in the real world.

Textbooks

Reading should be an integral part of any student's education in the life sciences. Primary literature often is inaccessible to students because of the level of scientific sophistication needed to understand the material, and popular literature often contains factual errors or biases that distort the information presented to readers. Therefore, textbooks are, and probably will remain, important vehicles for the presentation of biological concepts to students and should be designed to help students succeed in a biology program.

The current design and use of textbooks, however, lead us to recommend that: **All biology programs should use textbooks in ways that complement the active involvement of students and that support the development of structural and multidimensional levels of biological literacy.** To accomplish this, textbooks should use unifying principles to organize biological content and should include societal and historical issues to supplement textual material. Textbooks should be well-illustrated and visually stimulating. They should be functional, that is, easy for students to find information in and take notes from, and easy for publishers to change and update. Textbooks should have an appropriate reading level and a fluid reading style, and, importantly, textbooks should be compact.

Lectures

Many biology courses in colleges, universities, and high schools are taught using a lecture format to present students with a large body of information in a short time (Weiss, 1978; 1987). A lecture given by a dynamic, informed, and responsive individual can be a fascinating and worthwhile experience for any student. A judicious use of lectures is an effective way to transfer information from one individual to many, but **no biology course should be taught using lecture alone**.

The most appropriate uses of a lecture might be to pique the curiosity and wonder of students about biology, to provide alternative or expanded explanations of complex phenomena presented in the textbook, laboratory, or via educational technology, and to clarify, summarize, and reinforce the learning of concepts *after* students have discovered them through classroom investigations. Suggestions for improving lectures are presented in Chapter 6.

Learning and Teaching About Biology

Learning is the most important outcome of a curriculum and instruction. While focusing on learning, instructors and curriculum developers must address the needs of students who have diverse backgrounds and interests. How can one develop an effective biology program for a group of diverse students, and what must be included in a program to bring about learning? To answer these questions adequately, one must understand the learning process and consistently use an appropriate instructional model in a biology class. One also must select appropriate biological knowledge, curriculum themes, instructional strategies, and assessment instruments to use in class.

We consider each of these components of a biology program after first introducing an instructional model.

An Instructional Model

In recent years cognitive psychologists have shed some light on the learning process and proposed a model that may accommodate all students. The model is referred to as *constructivism*, a term that expresses a dynamic and interactive conception of human learning. Students redefine, reorganize, elaborate, and change their initial concepts through interaction between themselves and their environment and other individuals. The learner interprets objects and phenomena and internalizes the interpretation in terms of previous experiences. Such changes often require the instructor first to challenge the students' conceptions by showing these conceptions to be inadequate. To avoid leaving students with an overall sense of inadequacy, however, there must be time and experiences to reconstruct a more adequate conception than the original one.

The learning and teaching framework based on constructivism consists of five basic phases, summarized in Table 4-1. The first phase is to *engage* the students in the learning task. Once the students are engaged, they need time to *explore* the ideas through common, concrete experiences on which they can build concepts, processes, and skills. Next, the students must

Engagement
The engagement phase initiates learning. The activity should (1) make connections between past and present learning experiences, and (2) organize students' thinking toward the learning outcomes of current activities.
Exploration
The exploration phase of the teaching model provides students with a common base of experience within which current concepts, processes, and skills are identified and developed.
Explanation
The explanation phase focuses students' attention on particular aspects of their engagement and exploration experiences and provides opportunities to demonstrate their conceptual understanding, process skills, or behaviors. This phase also provides opportunities for teachers to introduce concepts, processes, or skills.
Elaboration
The elaboration phase challenges and extends the students' conceptual understanding and skills. Through new experiences, the students develop deeper and broader understanding and adequate investigative skills.
Evaluation
The evaluation phase encourages students to assess their understanding and abilities and provides opportunities for teachers to evaluate student progress toward achieving the educational objectives.

Table 4-1: An instructional model

explain—develop common explanations for their explorations. They then should *elaborate* the concepts, processes, or skills that have been explained. Finally, the students need *evaluation*—feedback about the adequacy of their explanations.

The instructional model proposed in this guide is a modification and expansion of the original learning cycle used in the Science Curriculum Improvement Study (SCIS) program. Three factors justify this instructional framework: (1) research from the cognitive sciences dealing primarily with student misconceptions, (2) concordance of the model with the processes of science, and (3) the model's utility for curriculum developers, biology teachers, and college professors.

The following sections describe the five phases of the instructional model. College instructors and classroom teachers can use the model to organize a year-long or semester-long program, units of the curriculum, or single lectures and lessons.

Engagement. The first phase is to engage students in the learning task. Students mentally focus on an object, problem, situation, or event. The activities of this phase make connections to past and future activities. The connections depend on the learning task and may be conceptual, procedural, or behavioral. Asking a question, defining a problem, showing a discrepant event, and acting out a problematic situation are a few ways to engage students and focus them on the instructional task. The role of the teacher is to present the situation and identify the instructional task.

Successful engagement results in students being intrigued by the activity and motivated to learn. Here, we use the word activity in both a constructivist and a behavioral sense, that is, students are both mentally and physically active.

Exploration. Once engaged by the activities, students need time to explore their ideas. Exploration activities are designed so students have common experiences upon which they continue building concepts, processes, and skills. This phase should be as concrete and hands-on as possible.

The aim of exploration activities is to establish experiences that teachers and students can use later to discuss concepts, processes, or skills. The teacher initiates the activity and allows the students time and opportunity to investigate objects, materials, and situations based on each student's ideas of the phenomena. As a result of their mental and physical involvement in the activity, the students establish relationships, observe patterns, identify variables, and question events.

If called upon, the teacher may guide students as they begin constructing (or reconstructing) their explanations. At both high school and college levels, the use of historical examples depicting the development of biological concepts serves as an excellent exploratory activity. The historical examples often align with and subsequently challenge the students' current explanation of concepts.

Explanation. The word *explanation* means the act or process in which concepts, processes, or skills become plain, comprehensible, and clear. Explanation, therefore, requires that students and teachers use common terms related to the learning task. In this phase, the instructor directs student attention to specific aspects of the two previous phases and formally introduces scientific explanations to bring order to the exploratory experiences.

The explanation phase is teacher- (or technology-) directed, and teachers have a variety of techniques and strategies at their disposal, including verbal explanations, videotapes, films, and educational software. This phase continues the process of mental ordering, and students ultimately should be able to explain in common terms their previous class experiences. The key to this phase is to present concepts, processes, or skills briefly, simply, clearly, and directly, and then to move on to the next phase.

Elaboration. Once students have developed explanations, it is important to involve them in further experiences that extend, or elaborate, the concepts, processes, or skills. In some cases, students still may have misconceptions, or they may understand a concept only in terms of the exploratory experience. This phase also involves students in new situations and problems that require the application of identical or similar explanations. Generalization of concepts, processes, and skills is the primary goal of elaboration.

Evaluation. At some point, it is important that students receive feedback on the adequacy of their explanations. Informal evaluation can occur from the beginning of the teaching sequence; formal evaluation can be completed after the elaboration phase. As a practical educational matter, teachers must assess educational outcomes. In this phase, teachers administer tests to determine each student's level of understanding and to allow students to evaluate their own understanding using their newly acquired skills.

The instructional framework proposed here parallels the actual processes involved in the scientific enterprise, and the methods of scientific inquiry are excellent tools for students to use as they test their explanations. How well, for example, do student explanations stand up to review by peers and teachers? Is there a need to reform ideas based on experience?

The remainder of this chapter deals with choices one may make about biological knowledge, curriculum themes, instructional strategies, and methods of assessment to develop a biology program that engages students and allows them to explore, explain, elaborate, and to be evaluated as they learn about biology.

Biological Knowledge

All biology programs consist of biological knowledge, but the form in which textbooks, software, or lectures present that knowledge varies considerably. Many biology textbooks are elaborate and colorful descriptions of thousands of biological facts; others emphasize major concepts or unifying principles of the discipline to organize biological information. The unifying principles used in this framework, and several major concepts related to each principle (Table 4-2), should help students and instructors organize the biological content of a course. For instance, we suggest avoiding a simple listing of definitions and functions of different cell parts,

EVOLUTION: PATTERNS OF CHANGE
Living Systems Change Through Time

- Forces of Evolutionary Change
- Patterns of Evolution
- Extinction
- Conservation Biology

EVOLUTION: PRODUCTS OF CHANGE
Evolution Has Produced Diverse Living Systems on Earth

- Specialization and Adaptation
- Species and Speciation
- Phylogenetic Classification
- Biodiversity

INTERACTION AND INTERDEPENDENCE
*Living Systems Interact with Their Environment and Are
Interdependent with Other Systems*

- Environmental Factors
- Population Ecology
- Ecosystems
- Biosphere

GENETIC CONTINUITY AND REPRODUCTION
*Living Systems are Related to Other Generations by Genetic Material
Passed on Through Reproduction*

- The Gene
- DNA (The Genetic Material)
- Gene Action
- Patterns of Inheritance
- Reproduction

GROWTH, DEVELOPMENT, AND DIFFERENTIATION
*Living Systems Grow, Develop, and Differentiate During Their Lifetimes
Based on a Genetic Plan That Is Influenced by the Environment*

- Patterns of Growth
- Patterns of Development
- Differentiation
- Form and Function

ENERGY, MATTER, AND ORGANIZATION
*Living Systems Are Complex and Highly Organized, and They Require
Energy and Matter to Maintain this Organization*

- Molecular Structure
- Hierarchy of Organization
- Matter
- Energy and Metabolism

MAINTENANCE OF A DYNAMIC EQUILIBRIUM
*Living Systems Maintain a Relatively Stable Internal Environment
Through Their Regulatory Mechanisms and Behavior*

- Detection of Environmental Stimuli
- Homeostasis
- Health and Disease
- Behavior

Table 4-2: Unifying principles of biology with several major concepts related to each

tissues of an organ, or growth and development hormones. Instead, we suggest using unifying principles to help students see the unity in structure and function within the diverse organisms on Earth. For example, when Evolution: Patterns and Products of Change is used as a unifying principle of biology, cell parts, tissues, and hormones may be seen as adaptations that have contributed to an organism's successful reproduction and to a species' reproductive success through evolutionary history. When teaching about cells (or tissues or hormones), students also can be shown how individual cell parts contribute to the organization of the whole by referring to the ideas set forth in the unifying principle of *Energy, Matter, and Organization*.

We recommend that individuals responsible for designing biology programs use these unifying principles of biology, which were introduced in Chapter 3 and are elaborated in Chapter 5. These principles represent a comprehensive foundation for the biological sciences and thus are common to high school, two-year, and four-year college programs. Individual programs may emphasize a molecular, organismic, or ecological level of organization; they may give greater emphasis to inquiry or societal problems, but the unifying principles should provide the foundation for the biological knowledge in all contemporary biology programs.

Curriculum Themes

A curriculum theme is an instructional orientation that helps organize and deliver a biology program to students. It is a systematic and deliberate expression of a viewpoint about the role of science subject matter, a coherent set of messages to the student *about* science (Roberts, 1982). The curriculum theme serves as a paradigm or frame of reference to help organize the teaching of biology. Whereas unifying principles help organize instruction of the *facts* of biology, a curriculum theme helps organize the structure of the biology *program*. For example, because evolution is the central theory of biology, it serves as a unifying principle that unites all biological facts. Evolution, however, also may serve as a curriculum theme that helps one select the activities of a class. With evolution as a *curriculum theme*, an instructor may choose, for example, laboratories that deal with mutation, genetic variation, and natural selection, and discussions that focus on speciation and extinction. Table 4-3 identifies each of the curriculum themes and recommendations for their use in high school and college biology classes.

Alternatively, an instructor may elect to use *personal aspects of biology* as a curriculum theme. By using this theme, one can organize a course around those issues, concepts, and activities that are important to students as individuals, such as careers in the biological sciences, individual health issues, and the process and problems of human sexual reproduction. Other themes are the *nature of science*, which focuses on the basic processes of biology as a scientific discipline and helps students become competent in their use; *social and ethical aspects* of biology focusing on bioethical issues; *technology and its influence* on science and people; and the *historical development of biological ideas*. *Biological content* has been a traditional curriculum theme for many courses through the years, emphasizing

CURRICULUM THEMES	High School	2-Year College	4-Year College
Nature of Science/Science as a Process	✓✓✓	✓✓✓	✓✓✓
Unifying Principles	✓✓✓	✓✓✓	✓✓✓
Biological Knowledge	✓✓	✓✓	✓✓
Personal Aspects of Biology	✓✓✓	✓✓	✓
Social Aspects of Biology	✓✓	✓✓	✓✓
Ethical Aspects of Biology	✓✓	✓✓✓	✓✓✓
Technology	✓✓	✓✓	✓✓
Historical Development of Biological Ideas	✓✓	✓✓	✓

✓✓✓ = ESSENTIAL   ✓✓ = OPTIONAL   ✓ = NONESSENTIAL

Table 4-3: Curriculum themes with suggested emphases for secondary and post-secondary programs

the facts generated by scientific investigation. We recommend that *biological content* never be used in isolation or as the only curriculum theme of a biology program.

Instructors should consider organizing their biology program around one or two main curriculum themes. "Essential" themes are those that are recommended for organizing a course. "Optional" themes are those that are discretionary in nature; although they may be used in a program, their use should be weighed carefully by the instructor before inclusion. Finally, "nonessential" themes are those we do not recommend using. These three levels of recommendation are used throughout this guide for instructional strategies, assessment strategies, and for the biological concepts in Chapter 5.

Instructional Strategies

The activities of a biology class make up the instructional strategies, and biology educators first ought to consider varying the activities they use. The use of a variety of teaching strategies accommodates various student learning styles and provides opportunities that will help different students construct their understanding of biological concepts. Table 4-4 lists a variety of strategies and suggests the degree to which the strategies might be emphasized at different levels.

Inquiry-oriented activities should be central to biology education, developing in students greater independence as individual thinkers and investigators. Whatever teaching and learning strategies one uses, instructional

CURRICULUM THEMES Explained
Allows students to learn about the structure and function of science.
Uses major biological principles to organize the program.
Emphasizes biological concepts.
Focuses on the impact of biology on the lives of individuals, including careers.
Emphasizes issues related to science and society.
Features ethical aspects of scientific investigation and of personal choices involving scientific information.
Introduces students to the influence of technology on science and on the lives of people.
Clarifies the processes of science and the development of biological explanations at different times and in different cultures.

Table 4-3 Continued

materials and curricular activities should emphasize student understanding of both concepts and the scientific processes of investigation. Instructional strategies should lead naturally toward individual student research of biological questions.

Although laboratories must be part of any classroom experience in biology, students also need opportunities to read, write, and explain biological concepts. Instructors should consider students' initial understanding of biological concepts and provide ample time for students to think about, interpret, and incorporate new information and ideas, or to revise their old views. Students should have opportunities to discuss and debate biologically related issues that are important to them and to work with, and possibly develop, models and simulations. Although learning and thinking independently are important student attributes, working cooperatively in groups also is important. Small-group activities can provide a dynamic learning experience not possible for individual students.

Clearly, numerous instructional strategies can enhance learning, but most science teachers consistently use one dominant method, lecture (Weiss, 1987). Biology education could make a significant step toward reform by using a variety of strategies in any chapter, unit, or course.

LEARNING/TEACHING STRATEGIES			
	High School	2-Year College	4-Year College
Student-centered Activities	✓✓✓	✓✓✓	✓✓✓
Direct Experience Activities	✓✓✓	✓✓✓	✓✓✓
• Laboratories	✓✓✓	✓✓✓	✓✓✓
• Field Experiences	✓✓✓	✓✓	✓✓
• Inquiry Activities	✓✓✓	✓✓✓	✓✓✓
• Individual Research Projects	✓✓	✓✓	✓✓✓
• Issue-Centered Problems	✓✓✓	✓✓✓	✓✓✓
Group Activities	✓✓✓	✓✓	✓✓
• Discussion	✓✓✓	✓✓✓	✓✓✓
• Debates	✓✓	✓✓	✓✓
• Cooperative Learning	✓✓✓	✓✓	✓✓
Individual Activities	✓✓✓	✓✓✓	✓✓✓
• Student Writing	✓✓✓	✓✓✓	✓✓✓
• Student Reading	✓✓✓	✓✓✓	✓✓✓
• Student Speaking	✓✓	✓✓	✓✓
• Student Explanations of Concepts	✓✓✓	✓✓✓	✓✓✓
• Analysis of Data	✓✓✓	✓✓✓	✓✓✓

✓✓✓ = ESSENTIAL   ✓✓ = OPTIONAL   ✓ = NONESSENTIAL

Table 4-4: Learning/teaching strategies with suggested emphases for secondary and post-secondary programs

LEARNING/TEACHING STRATEGIES Explained

"Hands-on" activities in which students investigate a question in biology using materials and equipment in a laboratory or natural area.

Field studies in managed ecosystems (such as a park) or natural habitat. Field studies may be simple or complex, short or long, but they should involve the students in an authentic investigation of some biological phenomena.

Activities incorporating aspects of a scientific process of investigation--reflecting a philosophy that methods for learning science should be the same as methods for doing science. Based on a student's question and his or her development of an answer or a proposed explanation for a biological phenomenon.

Investigations in which students study something new to them. Students plan and complete a project in which they ask a question based on observations of some natural phenomenon, gather evidence, and propose an explanation. Students outline a detailed proposal of the project before they begin their investigation, which alerts students and instructors to potential problems.

Opportunities for students to apply their knowledge to real-life situations and problems. Students study a biologically related issue to develop an understanding of the scientific and societal aspects of the problem. Students may study how the natural world works, and how problems can be solved through human efforts.

Instructor facilitates discussion of concepts and ideas.

Students choose a local issue related to a scientific or biological problem and which point of view to argue.

Students with specific roles work together in a team to accomplish shared goals. Essential components include: positive interdependence, face-to-face interaction, individual accountability, small group and interpersonal skills, and group processing. (Johnson and Johnson, 1991)

Writing is incorporated into many activities, including essays on exams, short explanations of lecture material, and library or independent research projects.

Students read primary scientific literature or articles based on it, making reasonable judgments on the appropriateness of research methods and arguing points of view supported by information in the article.

Students present oral reports of independent or group research or library work.

Students explain and relate concepts to each other or the instructor; includes a variety of activities such as concept mapping.

Students examine data collected by other investigators. Students explain how data were collected and analyzed and determine if they agree with the interpretation of the results.

LEARNING/TEACHING STRATEGIES	High School	2-Year College	4-Year College
Technology-related Activities	✓✓	✓✓	✓✓
• Audiovisuals	✓✓	✓✓	✓✓
• Computer Simulations	✓✓	✓✓	✓✓
• Kits and Manipulatives	✓✓	✓✓	✓
Teacher-centered Activities	✓✓	✓✓	✓✓
Lectures	✓	✓	✓
Demonstrations	✓✓	✓✓	✓✓
✓✓✓ = ESSENTIAL ✓✓ = OPTIONAL ✓ = NONESSENTIAL			

Table 4-4 Continued

Assessment Strategies

Assessment should outline for students the expectations for learning science. Assessment instruments are most valuable when they focus on higher-order thinking skills, understanding and use of biological knowledge, and the demonstration of competence in a setting that is meaningful to students. Methods of assessment should measure more complex types of learning than those measured by objective examinations alone and should emphasize what students can do with the knowledge they do possess. For instance, assessment instruments could require students to express in clear statements their biological knowledge as well as to apply this knowledge to novel situations. The assessment of diverse students in a classroom may be addressed by using diverse methods, including nontraditional assessment instruments such as portfolios and student productions of simulations and models. Table 4-5 provides recommendations for the use of different assessment strategies; the emphasis varies across high school, two-year, and four-year colleges.

Assessment should be an ongoing process that begins by determining the information and skills students bring to the class and continues with documentation of their progress throughout the course. Students should have the opportunity to express themselves through a variety of methods (e.g., oral and pictorial reports). Assessment must be both relevant to the instructional objectives and reflective of the instructional process. For instance, if hands-on laboratory activities are used, then students need to be assessed on their laboratory skills as well as their content knowledge. If inquiry skills are important outcomes, then students should be given

LEARNING/TEACHING STRATEGIES Explained
Programs that are stopped at appropriate points with questions for discussion interjected by instructor.
Computer simulations of natural phenomena and laboratory experiences not possible in the classroom.
Kits from biological supply companies allow users to apply and use technology in the process of scientific inquiry.
Verbal explanations of biological concepts. Lectures can be a dynamic and efficient way of presenting information to students, but they never should be the only way students experience science.
Illustrations of natural phenomena and scientific investigations that are difficult for students to see or conduct during a regular class period.

assessments that require the use of these skills and not merely the identification of factual information. In other words, it is important to link assessment with instruction. Instructors must assess those aspects of the program they say they value.

The linking of assessment with instruction is the first step in creating more authentic assessments. The second step is to seek assessment strategies that reflect real-life situations, involve multifaceted tasks, and that require the integration of skills over an extended period of time. The critical issue is whether the assessment strategy is congruent with the level of instruction. Just as learning progresses through levels of proficiency, so too should the assessments increase in complexity.

In the BSCS model of biological literacy proposed in this guide, the type of assessment used reflects the level of competency expected. At the initial level (where the student has only a *nominal literacy* of biological concepts) the assessment effort could focus on establishing the depth of understanding and prior knowledge the student has when entering the class. Assessment strategies appropriate for this level of learning would be short-answer questions, justified multiple-choice items, essay responses, and manipulative tasks that emphasize general skills related to the processes of science. To assess *functional biological literacy*, students might be required to recall, apply, and justify information through multiple choice, short answers, and essays. More revealing information, however, can be acquired by means of assessment tasks that require students to use their laboratory skills, to apply their knowledge to resolve problems, or to identify issues relevant to their lives.

ASSESSMENT	High School	2-Year College	4-Year College
Written Examinations	✓✓	✓✓	✓✓
Multiple Choice Questions	✓✓	✓✓	✓✓
Essay Questions	✓✓	✓✓✓	✓✓✓
Constructed Response Questions	✓✓✓	✓✓	✓✓
Extended Written Work	✓✓	✓✓	✓✓
Research Papers	✓	✓	✓✓
Project and Laboratory Reports	✓✓	✓✓	✓✓✓
Self Assessment and Journals	✓✓	✓	✓
Student Activity Assessment	✓✓	✓✓	✓✓
Performance Assessment	✓✓✓	✓✓	✓✓
Laboratory Practicals	✓✓	✓✓	✓✓

✓✓✓ = ESSENTIAL   ✓✓ = OPTIONAL   ✓ = NONESSENTIAL

Table 4-5: Assessment strategies with suggested emphases for secondary and post-secondary programs

ASSESSMENT Explained

Convergent problems and questions to which a single correct answer is best. Appropriate technique for the assessment of recall, recognition, and comprehension of information. Can be used to assess students' ability to analyze and make inferences if test items are carefully constructed (Millman and Green, 1989).

Divergent problems and questions to which alternative answers may be acceptable based on supporting evidence the student provides (Haertel, 1991). Useful technique for the assessment of structural and multidimensional literacy. Although essays allow students to express their depth of understanding and ability to connect ideas, inadequate writing skills can confound scoring for content knowledge (Hudson, 1991).

Response item requiring students to produce or construct their own answers (Bennett et al., 1990; Wittrock and Baker, 1991). Useful for the assessment of both nominal and structural literacy. Provides insight into students' prior knowledge, misconceptions, and reasoning processes. Constructed response items may be used as part of a performance-based laboratory for gathering data, as well as part of a computer simulation. Inter-rater reliability for using standardized scoring rubrics are equal to or better than essay questions (U.S. Congress, 1992).

Students may develop a paper synthesizing the work and ideas of other investigators. A useful technique to assess functional and structural literacy. To assess structural and multidimensional literacy, it is appropriate to have students report on an independent research project of their own design.

Students focus on the presentation and interpretation of results of a laboratory investigation or independent project. This assessment technique is appropriate for all levels of student literacy. Assessment items should emphasize the use of inquiry-based activities rather than the use of illustration or confirmation type of activities (Pickering, 1985).

Students write or tape a retrospective account of a problem-solving experience based on prompted questions. Includes self-inventories that measure attitudes and beliefs (Lester and Kroll, 1990).

Appropriate assessment technique focusing on analysis and application—can the student actively develop an answer to a question that is from the real world or that parallels a problem there? These assessment tasks should require an extended period of time to complete. Tasks should be related to several unifying principles of biology and blend essential content and essential processes. Performance assessments range from open-ended problems requiring students to define a problem and determine a strategy for solving it, to hands-on laboratory experiments that allow students to demonstrate competency in experimental design or dexterity in equipment manipulation, to written action plans that address community issues (Stiggins and Bridgeford, 1986; Baron, 1991; NAEP, 1987; Perrone, 1991; U.S. Congress, 1992). The choice of tasks should allow students to develop alternative solutions and require the use of essential rather than tangential knowledge or skills (NSTA, 1991). Performance assessments may include a variety of other assessment instruments such as the following six:

Assessment focusing on practical skills.

ASSESSMENT	High School	2-Year College	4-Year College
Oral Reports/Oral Examinations	✓✓	✓	✓
Interviews with Students	✓	✓	✓
Research Projects	✓✓	✓✓	✓✓✓
Observations of Students	✓✓	✓	✓
Peer Group Assessment	✓✓	✓	✓
Multidimensional Evaluation	✓✓	✓✓	✓✓
Construction of Models and Simulations	✓✓	✓✓	✓✓
Portfolios	✓✓	✓	✓
✓✓✓ = ESSENTIAL ✓✓ = OPTIONAL ✓ = NONESSENTIAL			

Table 4-5 Continued

As students develop an understanding of the subject, they can explain key concepts in their own words and can identify and describe the significance of these concepts to other concepts. At this *structural literacy* level, students should experience assessment tasks that require applications of their knowledge to novel situations. "Design-an-experiment" type of performance tasks, long-term research projects, and written reports are appropriate assessment strategies at this level of instruction. At the *multidimensional literacy* level, students can recognize deficiencies in their knowledge and are aware of more global applications of their knowledge. Students at this level are able to connect many diverse ideas, and the assessment methods should reflect this skill. Appropriate assessment strategies might require students to develop and undertake a plan of action to resolve a community issue.

Assessment should match the level of learning and increase in complexity as students become more proficient. At all levels, students should be allowed to perform and exhibit their ability in a variety of tasks. The bottom line is that good assessment methods, like good instruction, have clearly focused objectives, use a variety of techniques, are relevant to the students' lives, and are responsive to individual differences.

ASSESSMENT Explained

Oral presentations of work resulting from an inquiry into a biological question or responses to questions from instructor or classmates.
Assessment of the strategies that students use to solve problems and what they think about while working on them.
Results from projects designed and conducted by students on an inquiry into a biological question.
Direct observations of student performance, skills, and understanding. Appropriate for both group and individual assessments. Provides insights into the learning process as well as the quantity and quality of student contributions to group efforts (Hein, 1991; U.S. Congress, 1992).
Assessment in which students evaluate each other's performances, responses, contributions, and understanding of concepts.
Long-term projects used to measure student understanding of concepts or unifying principles. Students' development of physical models and computer programs may be considered as part of assessment.
A collection of writings, projects, and products that a student considers to be the best representation of progress over a period of time. Each item selected for inclusion is the result of the students' reflection on the appropriateness of that item to represent their progress (Wolf, 1989; Collins, 1992). Portfolios are not archival depositories but a constantly changing collection of materials that documents student growth. Portfolios are an effective assessment technique for both structural and multidimensional literacy.

Conclusion

One can think about a contemporary biology program as a set of interacting matrices, with the many factors affecting the learning and teaching of biology forming the sides of each matrix. Figure 4-1 is the matrix that illustrates curriculum components. This set of interacting sides defines how students learn about and how instructors teach biology. The three sides summarize the curriculum themes that may be used to organize a class, and the assessment instruments and instructional strategies used to teach the class.

What should students know, value, and be able to do by the time they have successfully completed a biology program? Figure 4-2 illustrates some of the desirable outcomes for students completing a contemporary biology program. Knowledge is not limited to biological content; knowledge also should include a working understanding of how scientific investigations are conducted and how all the facts of biology are unified by the unifying principles. Many of the desired values are incorporated in the characterization of a biologically literate individual and the information he or she should possess. Among these values are openness, curiosity, and respect for logic. Higher-order reasoning is desirable in a student who has completed a biology program. This trait includes the ability to evaluate critically a problem for possible solutions or a scientific statement for possible biases, incorrect assumptions, or invalid applications of data collection or analysis.

There are many interdependent curriculum components to consider for use in a contemporary biology program. Although it is highly unlikely any one program would include all curriculum components illustrated in Figures 4-1 and 4-2, the ability of students to attain desirable outcomes depends on a meaningful and varied classroom experience.

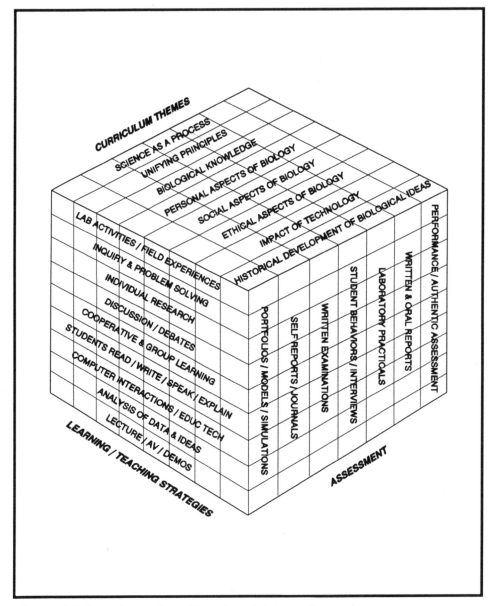

Figure 4-1: Learning and teaching about biology: Curriculum components

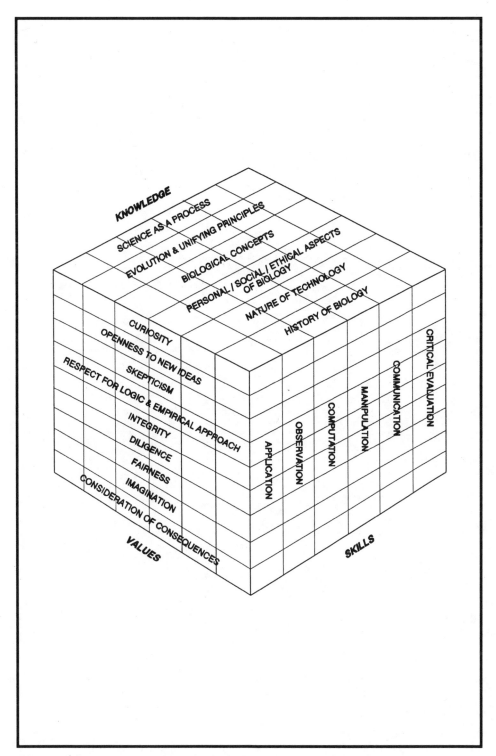

Figure 4-2: Learning and teaching about biology: Student outcomes

5

Unifying Principles of Biology

Biologists study living systems. There are millions of different species of organisms on Earth today, and one goal of biologists is to understand how these organisms differ from one another. Living systems, however, also share many characteristics, and the principles on which biology is founded apply to all organisms. These unifying principles form the basic structure for the study, teaching, and learning of the life sciences. The principles help to organize the subject and point to patterns and processes of natural phenomena in the living world, and they should play a central role in the design of any class, text, or program of study.

In this guide, we propose the following six unifying principles of biology: (1) Evolution—Patterns and Products of Change; (2) Interaction and Interdependence; (3) Genetic Continuity and Reproduction; (4) Growth, Development, and Differentiation; (5) Energy, Matter, and Organization; and (6) Maintenance of a Dynamic Equilibrium. These unifying principles have a well-documented foundation based on the research of many biologists throughout the history of scientific investigation. We used several criteria to select the principles:

- they reflect the most comprehensive, reliable information about living systems,
- they are well-supported by scientific investigations,
- they incorporate fundamental knowledge for understanding living systems,
- they demonstrate the underlying unity of living systems,
- they are characteristic of all living systems, and
- they are applicable to all levels of biological organization, from molecules to the biosphere.

Because evolution is central to the study and teaching of biology, we begin with a discussion about evolution and one clear manifestation of evolution—biodiversity. Following this discussion and two related essays, the six unifying principles of biology are identified and elaborated.

Evolution Is the Unifying Theory of Biology

How can one simultaneously account for the extraordinary diversity and observable unity of living systems in the world today? The answer, in a word, is evolution. Evolution is *the* unifying theory of biology because it has played a role in the history and lives of all living organisms on Earth today—and of those that are now extinct. Evolution is the major conceptual scheme of biology because it helps us understand relationships between organisms, past and present, and the many ways organisms have succeeded in different habitats.

First, consider the origin of life on Earth. The Earth is approximately five billion years old. Within a billion years after Earth's formation, living things evolved from a lifeless world. Evidence from a variety of sources suggests that before there was life on Earth, there were molecules similar to those found in living organisms today. These molecules were more organized than the materials from which they were made—organization requiring energy. With the energy sources and raw materials known to be present in the prelife Earth and a billion years of time, many prototypes of the first molecules of life formed and were destroyed by the physical environment. Eventually, a few of the molecules may have survived and reproduced themselves from the surrounding materials. These surviving molecules became prevalent and joined with others; inner molecules became protected from the environment by a naturally forming "membrane" of outer molecules. Some molecular combinations may have been able to survive longer than others, reproduce, and then become dominant in the lifeless world. Piece by piece, the first cells assembled themselves, but the evolution to reproducible organized units was a very slow process. These first systems were able to maintain their physical integrity, use materials and energy from their environment, respond to the environment, and reproduce similar systems; that is, they had characteristics of life.

Since that time, cells that reproduced cells similar to themselves have been able to survive on Earth, some growing larger by forming associations with other cells. A few of these novel aggregations colonized, survived, and reproduced in new habitats. There are, however, many habitats around the world, and many ways to survive in them. The variety of ways and places to survive accounts for the diversity of living systems on Earth today.

Biodiversity

There are millions of different species of organisms living on Earth today. This biodiversity reflects the many ways organisms obtain and use the resources required for life. From its beginnings in the ancestral pools of organic compounds, life has evolved, resulting in the great diversity of living forms and functions. Specialization of molecules, cells, tissues, and organs for different functions and different habitats accounts for much of the diversity of life. The first living things most closely resembled prokaryotic organisms----small, relatively simple cells. Today, biologists classify thousands of different prokaryotes inhabiting every part of the biosphere. Though microscopic in size, the numbers of prokaryotes are astronomical.

Eukaryotic organisms—protists, fungi, plants, and animals—also have evolved. Although not as numerous as prokaryotes in terms of individuals, these species dominate the Earth because of their size and use of resources, growing and reproducing in diverse environmental conditions. Eukaryotes differ from prokaryotes in several ways; most important, a eukaryotic cell contains organelles, each with a distinct function, and a prokaryotic cell lacks organelles.

The ability to adapt to widely differing environmental conditions also explains the variety of living organisms. The environment of an organism consists of living and nonliving factors. The combination of these biotic and abiotic factors varies from one part of the Earth to another. The variety of environments suggests that certain characteristics would be advantageous to an organism in one habitat, but might be less useful or even detrimental in a different part of the world. Through many generations, changes occurred, and advantageous characteristics were passed on to offspring so that organisms on Earth today show little resemblance to their ancestors or to related organisms in different habitats. In the following essay, Lynn Margulis and Dorion Sagan describe the interdependence of the diverse species on Earth and the importance of biodiversity to humans.

ESSAY

Biodiversity: A Matter of Species Survival

by Lynn Margulis and Dorion Sagan

We human beings are in the habit of considering ourselves above nature or apart from it. We like to think of ourselves as special—if not the chosen species, then the most accomplished one, whose brains, or toolmaking ability, or ability to make arches, or opposable thumbs, or erect stance, speech, or self-consciousness, make us superior to every other life form. Humans seem determined to find or declare something—anything—with which to prove ourselves the most highly evolved species on Earth.

This haughty attitude is part of our self-confidence as a species; it has served us well. As it can be healthy for an individual to believe strongly in his or her own self-worth, so anthropocentrism can promote the survival of a social species. The Bible says that plants and animals were created for the benefit and use of man; by ten thousand years ago people already were in the habit of growing plants in agriculture, using animals for clothing, meat, and moving. Today people are more aware that through our activities we are producing the extinction of many species. But the widespread belief in our specialness, anthropocentrism, has remained. The difference is that now, instead of thinking that our specialness countenances us to use other organisms as we see fit, we believe it is our duty to be in charge of, to manage, to engineer the make-up of life on Earth, to be planetary stewards. Instead of thinking living beings are God's gift to us, we now act as if we were the lords of the Earth. We still think we are special.

There is a thin line between self-confidence and megalomania. And this thin line moves: the anthropocentrism, the species self-centeredness that was once a survival trait has now become a burden to our survival. *Homo sapiens* began as an obscure African primate, but has now evolved, like a mammalian weed, into over 5 billion people inhabiting all land masses on the planet. Such wild success transforms ecological relationships. Long before plants and animals ever evolved, photosynthetic cyanobacteria mutated to use water as a source of hydrogen in photosynthesis. This innovation was as important for the cyanobacteria as technological intelligence was for humans. It helped the bacteria spread across land surfaces and through waterways all over the planet. But, as with *Staphylococcus*, a microbe normally found on skin that can get out of control and cause us infection, wild ecological success may have serious consequences. The cyanobacteria that used water as a source of their hydrogen produced oxygen as their waste, giving Earth for the first time an oxygen-rich atmosphere. This oxygen was similar to the pollution produced by our cars and factories. Indeed, although eventually it was successfully disposed of by many types of respiring organisms, oxygen killed most of the bacteria that produced it. Cyanobacteria caused world-wide pollution that pushed other gene-trading types of bacteria underground, out of the way of this pollution.

Organisms struggle but they also work together to survive. The human body contains so many normal bacteria, fungi, and other microbes that if all our animal cells were zapped away, our silhouette would still shimmer, a living ghost made of these subvisible organisms. Some of these microbes are benign, others, such as the bacteria in our gut that aid in the metabolic production of vitamin B_{12}, are positively helpful. Biochemical and genetic studies have proved beyond a reasonable doubt the thesis that nucleated cells comprising animal tissues are the result, evolutionarily, of symbiosis—the living together of diverse types of bacteria. Different bacteria that formed alliances, billions of years ago, are at the core of our being. No ecosystem of one single or only a few species can persist; biodiversity is a prerequisite to survival.

As went the past, so goes the future. To live on a planet we have crowded with our toxins and our selves requires that we forego the selfishness that served us so well in our phase of settling throughout the Earth. Approximately thirty million species now co-inhabit the Earth. We are one. Yet because of poaching, clearing land for monocrop agriculture, razing the rainforest for cattle and cars, and other activities typical of technological humanity, animal, plant, and microbial species of little direct interest to urban people become extinct daily. Indeed, the paleontologists have warned that the present rate of species extinction at the hands of humanity is the greatest since the Cretaceous Era when not only the dinosaurs but many marine microbes, land plants, and other species disappeared forever. Species today suffer extinction before they can even be named, recorded, and described in the scientific literature!

Biodiversity is the word used to discuss the richness of different life forms living together. All animals, plants, fungi, and protoctists on the planet have descended from biodiverse aggregates, from bacteria inhabiting the cytoplasm of other different bacteria. Biodiverse bacterial aggregates first formed new cells, protoctist cells, from which all animals, plants, and fungi later evolved. A recent taxonomic grouping of the bewildering diversity of 30 million living species into the five kingdoms is depicted in the table at the end of this essay.

There is strength in numbers, in the pioneering stage when members of a newly evolved species rapidly spread into, as yet, unexploited environments. But to survive after the period of spread requires reconciliation and integration in the new environments. No organism is an island. The elements making up life—carbon, nitrogen, oxygen, hydrogen, phosphorus, and sulfur—are limited on Earth and must be efficiently recycled. No single organism, or even single population of organisms, can recycle the elements on which it depends. Certain collections of bacteria, living together in organized communities (such as microbial mats) are metabolically diverse enough to provide for themselves. But all so-called "higher" organisms (and our view of ourselves as the highest is just another form of that old self-aggrandizement) depend on the gas-exchanging, sewage-transforming, pollution-dispersing skills of the microbial world: the bacteria, protoctists, and fungi. Wherever there is a living human or a tiger or even a dung beetle there is an entire supporting biosphere. The biosphere—the system of Earth life—is by definition biodiverse enough to be self-maintained. Yet, although bacteria, like those forming mat communities, can accomplish total recycling of the elements listed above by themselves, recycling occurs more quickly and thoroughly when more complex organisms such as large seaweeds and fish are present in the ecosystem.

Certain natural settings rich in vegetation and animal life often induce in us a sublime feeling of serenity and beauty. The green environment, uncrowded with other humans, whether cloud forest or savannah, sea coast or mountain valley, awakens something deep inside us. This feeling is probably no accident. "Biophilia" is the name scientists have given to our natural attraction toward other life forms—toward nonhuman nature. Whatever our anthropocentric delusions of grandeur, people were never self-sufficient. We are the children of tropical forests, of the bright-colored fruit and sweet-smelling flowers that attracted and fed our primate ancestors. We are the children not only of our own primate ancestors but of their biodiverse environments, rich in fauna and flora, fungi, and other microbiota. When our ancestors turned to hunting and eating meat to live, they depended not only on animals such as cervids (venison), bovines (buffalo hide, milk, cheese) and gallinaceous birds (chicken, turkey) but on the grasses (corn, wheat, rice) and other seed plants in the soil of their underlying environments, all teeming with life.

We should protect Earth's biodiversity not only because the rainforests may harbor the medicines or wonder hormones of tomorrow. A far deeper rationale is the one confirmed by the inner feeling of natural awe that overcomes us upon visiting a biodiverse scene: the interliving of distinct species is not just some duty in our charge as self-appointed stewards of Earth, it is the birthright of those organisms that, beyond their initial wild phase of growth, continue to survive.

Superkingdom (Domain)	Bacterial cell organization (chromonema, small ribosomes, continuous DNA synthesis). All metabolic modes represented.
Kingdom Monera	Single homologous genomes
• Archaebacteria	Archaebacterial ribosomes and 16S rRNA; ether-linked lipids (examples: methanogenic, thermoacidophilic, and halophilic bacteria)
• Eubacteria	Eubacterial 16S rRNA and lipids; peptidyl-glycan cell walls (examples: enterobacteria, phototrophic bacteria including cyanobacteria*, spirochetes, and other Gram-negative bacteria; Gram-positive bacteria; mitochondria and plastids)

Table 5-1: Five kingdom classification of life   *formerly blue-green algae

Superkingdom (Domain)	Nucleated cell organization with chromatin (histones, nucleosomes, nuclear pore complexes); large ribosomes; mitotic karyokinesis; actin-based cytokinesis; microtubule-based. intracellular motility systems; intermittent DNA synthesis
Kingdom Protoctista	Microorganisms and their larger descendants composed of multiple heterologous genomes (examples: photoautotrophic algae, osmohetero-trophic slime molds, phagoheterotrophic ciliates)
Kingdom Fungi	Haploid or dikaryotic, osmotrophic heterotrophs that develop from fungal spores and display zygotic meiosis. Chitinous cell walls (examples: yeasts, mushrooms, lichens)
Kingdom Plantae	Gametophytes develop from haploid spores; fertilization produces sporophyte embryos that are retained in maternal tissue. Cellulosic cell walls (examples: bryophytes, tracheophytes). Photoautotrophs.
Kingdom Animalia	Diploids develop from products of anisogamous (sperm/egg) fertilization into blastula embryos (examples: porifera, mollusks, chordates). Phagotrophic heterotrophs.

Table 5-1 Continued

Unity in Diversity

Although organisms are diverse, they also share many characteristics. First, all living organisms possess certain attributes that characterize life. Living systems: (1) change through time; (2) interact with their environment and are interdependent with other living systems; (3) grow, develop, and differentiate during their lifetimes; (4) have genetic continuity with other generations through reproduction; (5) are highly organized and use energy and matter to maintain this organization; and (6) maintain an internal dynamic equilibrium. These six characteristics of living systems form the bases for the *Unifying Principles of Biology*.

Although organisms may look and behave differently from one another, the unity among life forms becomes apparent if one analyzes their internal components. All organisms are composed of the same four kinds of carb-on-based biological macromolecules. These molecules vary in proportion and composition among different species. The genetic basis for inheritance of information is also the same in all known organisms. All living organisms possess DNA, which serves as the carrier of genetic information passed from one generation to the next as well as the carrier of information for the processes of daily life.

All organisms are linked to their ancestors through evolution, and scientists classify life on Earth into groups related by ancestry. This phylo-genetic classification scheme is based on the understanding that related organisms descended from a common ancestor and thus possess certain similar characteristics. Evidence of common ancestry includes the similarity in certain biochemical processes and physical features as well as the universality of the genetic code and macromolecules.

The many similarities among organisms provide clues to an ancient, common ancestor. Although many differences among organisms have accumulated through time, there are enough similarities to allow us to conclude that organisms on Earth today have a common heritage united through their evolution. In the following essay, Ernst Mayr describes the historical development and major components of current evolutionary theory.

ESSAY

Evolution

by Ernst Mayr

Among students of nature and philosophers from the Greeks on, there were essentially three concepts of the world: (1) an eternal, either constant or cycling, world; (2) a recently created, unchanging world, as was believed by Christians and described in Genesis; and (3) a world of long duration, continuously changing, and perhaps progressing toward a final stage of perfection.

The discoveries of the 17th and 18th centuries, particularly in cosmology, geology, and natural history, increasingly provided evidence for the validity of the third of these concepts, but it was not until shortly after 1800 that Lamarck published the first comprehensive theory of evolution. The general acceptance of the concept by scientists and informed members of the public followed the publication in 1859 of Darwin's *On the Origin of Species*.

Three different modes of evolution have been proposed. According to *saltational evolution*, a species may once in a while produce a uniquely different individual (a "mutation") that becomes the progenitor of a new species or higher taxon. This mode of evolution was proposed by typologists for whom each species was a different type separated by a gap from all others. According to *transformational evolution*, a given object or entity gradually changes over time. Indeed the word *evolution* was first used for the ontogenetic development of an organism from the fertilized egg to the adult stage. Lamarck's theory of evolution was one of transformation; he postulated the spontaneous origin of very simple organisms and their gradual subsequent transformation into higher and higher types ultimately culminating in the origin of humans. *Variational evolution* was Darwin's novel concept of evolution, postulating the production in every generation of an enormous number of variable individuals within every species, the subsequent survival of a few of them, and the production of a new highly variable population as the result of the interbreeding of the few survivors of the previous generation.

Modern biologists consider evolution as much a fact as that the Earth circles the sun. The evidence for evolution consists of the fossil record in which objectively dated (by radioactive methods) geological strata often yield completely continuous series of fossil forms that become more similar to living types the younger the fossils are. Further factual evidence is produced by embryology where, for instance, the descent of the terrestrial tetrapods from fishlike ancestors is revealed by the embryonic development of gill-arches. Embryos very often reveal such manifestations of the past history of their lineage. The molecules of which an organism consists have undergone an evolution during their history, and this evolution now can be reconstructed. This molecular history is consistent with that derived from other evidence (morphology, embryology, and geographic distribution).

Evolution consists of two processes. First, it is a change in the time dimension resulting in the maintenance or even improvement of adaptedness. One might consider this as "vertical evolution." Evolution, however, is also a process involving populations. Because each species consists of numerous populations, including those isolated beyond the periphery of the contiguous species' range, evolution also can be considered a "horizontal" process. Here, evolution leads to ever-greater diversity of the populations of a species, and eventually to the production of populations that can be considered incipient species, ultimately resulting in speciation. Hence, the two dominant processes of evolution are the production of adaptedness and the production of diversity.

One of Darwin's most brilliant insights was the realization that closely related species are descendants of a common ancestor. All cats, including lions, tigers, and cheetahs, are ultimately derived from one original cat ancestor. And this cat ancestor and the equivalent stem ancestor of dogs, bears, weasels, and hyenas are derived from a common carnivore ancestor. By tracing the ancestry further and further back, Darwin was able to suggest that not only animals and plants had been derived from a common ancestor, but indeed all of life on Earth. The discovery by molecular biology that all organisms, even bacteria, have the same genetic code, is powerful evidence supporting the validity of this theory.

The research that has as its objective the determination of the common ancestors of all organisms is called *phylogeny*. Even where there seemed to be gaps between major groups of organisms, such as between birds and reptiles or amphibians and fishes, certain fossils such as *Archaeopteryx* and *Ichthyostega* happily turned out to be perfect links. The discovery of such "missing links" has been particularly important for the reconstruction of the ancestry of humans; fossil hominids such as *Homo erectus (Pithecanthropus)*, *Homo habilis*, and *Australopithecus africanus* and *A. afarensis* have provided an almost perfect connection between modern humans and the chimpanzee.

What causes the evolutionary change from generation to generation? To answer this question there have been numerous theories from Lamarck on, but all of them, except for Darwin's theory of natural selection, were refuted by about 1940. What were the major refuted theories? One refuted theory was *saltationism*, which postulated that a single major genetic mutation would produce a new kind of individual that then gave rise in its offspring to a new species and possibly a new higher category. This kind of evolution does indeed occur in exceptional cases, such as in polyploidy and in the evolution of asexual forms of life. But it is now well understood that the seeming saltations from one type of organism to another, as reflected by the gaps in the fossil record, are not due to such saltations but rather are the result of extinction and incompletenesses of the fossil record. Darwinian evolution in sexually reproducing organisms takes place through the gradual change of populations and is therefore, by necessity, gradual.

Also refuted were *teleological theories* that stated evolution was due to an intrinsic capacity of phyletic lineages to evolve toward an ever-greater perfection, as Lamarck had thought. These theories had to be abandoned because no mechanism was found that would generate intrinsically such a trend toward perfection. Furthermore, the careful study of phyletic lineages revealed many reversals and, overall, an absence of the postulated rectilinearity.

Environmental causation, which proposed that organisms improved through use of their needed structures and functions and that this improvement was conveyed to future generations, also was refuted. It was postulated that often there was a direct influence of the environment on the genotype. Such theories were particularly favored by neo-Lamarckians. However, the rise of genetics showed that the genetic material is constant and cannot be modified by use or disuse or through direct induction by the environment. The demonstration by molecular biology that no information can be transferred from the phenotype to the genotype, that is from the proteins to the nucleic acids, provided the final refutation of all such environmental theories.

The refutation of all non-Darwinian theories of evolution left natural selection as the sole occupant of the field. However, because there were many uncertainties as to the nature and action of natural selection, this victory of natural selection was not the end of the controversies. The term *selection*, chosen by Darwin in analogy to artificial selection in domestic productions, was somewhat unfortunate. The term induced the query "who selects?" When Darwin said "nature," he was blamed by substituting nature for God without basically changing the causal chain.

We now realize that natural selection is an *a posteriori* phenomenon. Every set of parents produces numerous offspring, often thousands of them or even millions, but on the average only two of them remain to perpetuate the lineage. Those that are successful in survival owe this both to their superior constitution (statistically) as well as to luck because many individuals of all

species either die or are otherwise prevented from reproduction by accidental causes such as volcanic eruptions, weather catastrophes, and other singular events. Selection thus is a probabilistic process, describing the probability that those individuals best adapted to the current environmental conditions will have the greatest probability of survival, but also admitting that survival and reproduction will be affected by many contingencies.

There has been much controversy in recent years about the "unit of selection." The question is, what is the real target of selection? Is it the gene, the individual, the group to which an organism belongs, or its species? For Darwin (in the pre-genetic days) it was clearly the individual, and this is still the choice of most naturalists. An individual, however, may be successful in survival owing to the possession of a particular gene. It is important therefore to make a distinction between selection *of* and selection *for*. The question after the selection *of* is almost invariably answered by "an individual." The selection *for* may be a gene or a whole part of the genotype.

Is there group selection? Do species have certain characteristics that are favored by selection, not of an individual but only of a group? One can speak of group selection only if the group has a selective value that is greater than the arithmetic mean of all the individuals of which the group consists. This occurs only in social species or, in the case of humans, species with a culture. In such species there may be individuals that contribute altruistically to the welfare of the group even though this is disadvantageous to their own survival and reproduction.

There is a continuous replacement of species in the course of evolution caused by the extinction of species and evolutionary success of new species. This process has been referred to as *species selection*, but I prefer to use the term species turnover or species replacement because the actual mechanism is the elimination of individuals of one species and the evolutionary success of individuals of the other species—in other words, selection among individuals.

The fitness of an individual is measured not only in its survival but also in its reproductive success. Many of the attributes of an individual that lead to reproductive success do not necessarily have a survival value. The train of the male peacock or the gigantic size of elephant seal bulls are well known illustrations. All cases in which individuals have superior reproductive success either owing to combat among males or to female choice were combined by Darwin under the term *sexual selection*. Biologists now know that the domain of reproductive success is larger than mere sexual selection because it plays a role in many other aspects of life history, such as sib rivalry, age at reproduction, and parental investment.

Darwin never solved the puzzle of the origin of variation, a puzzle finally solved by genetics. First of all, there is a constant recurrence of spontaneous new mutations, many of which are retained in the gene pool in recessive condition. More importantly, the recombination of genes produces ever-new genotypes, each of which has its own unique survival value. In sexually reproducing species, no two individuals are genetically identical—in other words, there is an almost unlimited amount of variation available.

In the days of Darwin, Pasteur and others had demonstrated the failure of the occurrence of spontaneous generation ("new life") under the present environmental conditions. As a result, one of the arguments against Darwin's theory of evolution was that it failed to demonstrate the origin of life. Since then, geologists have demonstrated that the Earth in its earlier periods lacked oxygen and that any new life then originating was not immediately destroyed by oxygen.

There are a number of scenarios explaining the transition from inert molecules to life. It probably never will be possible to achieve complete certainty concerning these steps because they are not preserved in the fossil record and appear to be equally probable. The origin of life, however, is no longer the great puzzle it was before the development of modern chemistry and molecular biology.

The origin of life does not explain automatically how the enormous diversity of living organisms originated. Darwin realized that species were not constant and not only changed in time but also split into several species. Such a multiplication of species is called *speciation*. Even though instantaneous speciation occurs occasionally through polyploidy, the normal process of speciation is *geographic speciation*. In this process a population of a species becomes isolated from the remainder of the species by a geographical or ecological barrier and acquires cross sterility or other isolating mechanisms, so that after the removal of this barrier the two previously isolated populations now can invade each other's ranges without interbreeding. In another mode, a new founder population is established beyond the previous range of the species, and this isolated population in due time evolves into a separate species. It is still controversial to what extent *sympatric speciation* also occurs, that is, by the establishment of an ecologically specialized population within the breeding range of the parental population.

Speciation in asexually reproducing organisms is quite different. Each clone of such organisms is reproductively isolated from the other clones, and they become progressively more different from each other, owing to mutation and selection. Natural selection favors certain clones while others become extinct, and this produces gaps between these asexual clones, which eventually become as different genetically as sexual species.

There was a long argument after the publication of the *Origin* about whether the genetic changes within populations and the production of new species were able to explain so-called *macroevolution*, that is, the origin of new types of organisms. There is now rather general consensus that as far as the genetics of macroevolution is concerned, it is simply a continuation and elaboration of the processes that occur within species and during speciation. Higher taxa develop when species enter new adaptive zones, such as when the first bats became flying mammals and the first whales became obligatory inhabitants of the oceans. Ultimately, macroevolution occurs through the acquisition of new adaptations by populations and the concurrent genetic turnover.

Evolutionary novelties can occur either by a gradual modification of an existing structure, such as the modification of the legs of mammals not just for running but also for swimming or digging, or the modification of the anterior extremities of both bats and birds into wings. Another possibility is the adoption of a completely new function such as in fish swim bladders derived from lungs, and the conversion of the antennae of Cladocerans (*Daphnia*) into a swimming paddle. This latter process of a change of function explains how natural selection can contribute to the evolution of organs before they have adopted the new function, as in the conversion of an anterior extremity into a wing. This is totally overlooked by those who ask, how could natural selection have contributed to the evolution of the wing before birds were able to fly?

Spectacular aspects of macroevolution are the gradual development from the most primitive bacteria to one-celled organisms (protists) to many-celled higher eukaryotes such as fungi, plants, and animals, as well as the progressive development of such features as endothermy, a large central nervous system, and the adoption of parental care permitting transfer of cultural information from generation to generation. Evolutionists have not hesitated to refer to this sequence of changes in the history of the Earth as *evolutionary progress*. If one uses this terminology, one must realize that this seemingly upward-directed trend was not in any way built into the organic system, nor was it programmed in any way. The trend was simply the result of continuous high variation and continuing competition among individuals, populations, and species. Parallel to the evolution of new types of organisms is an evolution of the biota. The first 2 billion years of life on Earth were dominated by prokaryotes. Multicellular animals appeared rather suddenly in the late pre-Cambrian some 700 million years ago, all of them being aquatic. This fauna was dominated by such extinct types as the trilobites. After plants had invaded the land, various types of animals also became terrestrial, particularly arthropods and vertebrates. The Mesozoic was dominated by reptiles, which in turn gave rise to mammals and birds. Jurassic and Cretaceous life was dominated by the dinosaurs; and the coexisting mammals, which were small and probably nocturnal, did not have a major outburst of evolution until after the extinction of the dinosaurs at the end of the Cretaceous. Birds, likewise, particularly the smaller songbirds, seem to have had their radiation in the Tertiary period, that is, during the last 60 million years.

There is a steady turnover of species—some become extinct while other new ones originate. In addition, there are occasional periods of mass extinction including the one at the end of the Permian and the one at the end of the Cretaceous. There is general agreement that this last major extinction coincided with (and was apparently caused by) an asteroid impact on the Earth. It is uncertain and indeed rather improbable that the other mass extinctions, scores of which have been described by the paleontologists, were likewise caused by cosmic objects. Rather, it seems that strong fluctuations of the temperature of the Earth, as well as various consequences of plate tectonics were responsible for the earlier extinctions.

Under the influence of the theories of the geologist Lyell (uniformitarianism) and in line with the earlier evolutionary theories (transformationism), it was long assumed that evolution proceeded at a rather constant rate. That this is not the case was already known to Darwin, but was generally ignored. Opposed to this theory of constant evolutionary rates is that of speciational evolution (*punctuated equilibria*), according to which there may be a considerable speeding-up of evolutionary change during speciation, particularly in founder populations, while well-established species may sometimes show no visible change for many millions of years. As in many other evolutionary phenomena, there is apparently great variety with respect to rate of evolution. Even some founder populations do not undergo major changes, while some species apparently have shown steady changes without any multiplication of species. It will require much further research to determine how frequent the various possibilities actually are.

In the post-Darwinian period there was much disagreement among evolutionists, particularly concerning the validity of Darwin's theory of natural selection and the relative frequency of saltational or Lamarckian processes. This period of uncertainty was terminated by the so-called Evolutionary Synthesis of the 1930s-40s, in which the now broadly accepted Darwinian paradigm was agreed on. Since about 1970 there have been frequent attacks on this neo-Darwinian paradigm, but it appears that nothing was found that is in real conflict with the Darwinian picture. Even the findings of molecular biology have not led to any need for revision. To be sure, molecular mechanisms have been discovered in addition to mutation and recombination that would lead to increased variability and various forms of genetic recombinations previously unknown. However, what all these mechanisms do is to increase the variability of populations, with natural selection invariably entering the picture and determining the ultimate outcome. This is also true for so-called neutral evolution, that is the replacement of nucleic acid base pairs and of amino acids by such others that do not affect the fitness of the resulting phenotypes. Such "neutral" changes, because they do not affect the phenotype, are really only evolutionary "noise," and do not contradict in any way the basic Darwinian mechanism.

Everything we have learned in the last 100 years about the evolution of humans and their ancestors fits neatly into the picture provided by the Darwinian theory. It is now clear that the lineage of hominids leading to humans arose out of the group of African apes; indeed the chimpanzee in its molecules is more closely related to humans than the other African ape, the gorilla. The early hominid ancestors all were restricted to Africa, and it is only a little more than one million years ago that *Homo erectus* invaded Europe and Asia.

The various typically human characteristics did not originate simultaneously. Bipedal walking already had been acquired by *Australopithecus africanus* about 3 million years ago, while the most drastic increase in brain size, characterizing Neanderthals and modern humans, started only about 350,000 years ago. As one anthropologist has said recently, everything in the evolution of humans has happened far more recently than we formerly thought.

In spite of all we have learned about evolution, including the evolution of humans, there is a great deal still to be learned, and new discoveries are made every year. It is very gratifying for the student of evolution that none of the numerous recent discoveries has necessitated any major revision of the basic Darwinian paradigm as it was achieved during the Evolutionary Synthesis.

The Unifying Principles

Many reports and textbooks have identified unifying principles of biology and, although stated and organized differently, the major concepts in biology are contained in all such listings. Some lists incorporate more major concepts than others, but biology teachers and curriculum developers should emphasize unifying principles in their programs because the principles help students make sense of the myriad biological facts and help explain living systems.

The following statements provide an introduction to each of the unifying principles in a context appropriate for biology education. The statements are intended neither to be all-inclusive nor separate from each other; there is overlap between different principles because they are related. The introductions are based on statements developed by BSCS several decades ago and on the work of individuals involved in this project. Following each introduction is a short collection of research questions currently being addressed by scientists around the world. Also associated with each introduction is an essay (or two) that provides a contemporary view of the unifying principles from the perspective of a biologist recognized for research in a related field.

We have identified major concepts related to each unifying principle and given a recommendation for using each concept in a biology program at the secondary or post-secondary level: (a) *essential*—those concepts that should be included in a program; (b) *optional*—those concepts whose use should be weighed very carefully before inclusion, and (c) *nonessential*—those concepts whose use is not recommended.

The lists of concepts should never be used as an argument of what to "cover" in a course. Rather, the lists can be used to set priorities of what to *emphasize* and what to *eliminate* in a biology program. The lists also illustrate how the unifying principles of biology can be used as overarching constructs to organize whatever content *is* emphasized in a course.

PRINCIPLE 1. *Evolution: Patterns and Products of Change*
Living Systems Change Through Time

This principle of *Patterns and Products of Change* contains the major concepts of evolution as well as other causes and consequences of change. One aspect of this unifying principle is the change in the number and genetic constitution of individuals within a population through time. In the biological world, long-term success is related to the number of surviving offspring an organism produces: the more offspring that survive and reproduce, the more successful is the organism. To reproduce, organisms require resources such as energy, nutrients, and water, so reproductive success depends on the acquisition of resources that allow an organism first to grow and develop, and then to reproduce. The physical characteristics used to acquire resources are determined by genetic information; genes direct the development of structures that help an organism obtain and use the raw materials of life that are in limited supply. The fact that individuals of most populations produce more offspring than survive results in competition for the finite resources necessary for living. Within a population, however, organisms are not identical. Each member of a population may possess a unique complement of genes, which can lead to the expression of different phenotypes in the population. Those individuals better suited to acquire necessities are able to produce a greater number of offspring and thus pass their genes on to a greater proportion of the next generation. This genetic variation is the basis for evolution and is the major focus in the study of population genetics.

Organisms live within an environment that is constantly changing. Those individuals with little tolerance for environmental fluctuations may fail to reproduce. Thus, their unique combination of genes will not be passed on to any offspring. A species of organisms can become extinct with a resultant loss of its entire genetic complement, as has happened many times since the origin of life on Earth.

Natural selection is the process whereby individuals best adapted to their environment survive and reproduce successfully. Through time, the characteristics of individuals within a population will change as natural selection acts on the genetic variation within that population. This change in the frequency of alleles within a population through time is evolution. To appreciate the field of biology, one must comprehend the difference in scale when considering the size of the universe and the atoms composing it, and to appreciate evolution as a process, one must comprehend the length of time the Earth and life on it have existed. A billion years is difficult for most people to comprehend—and a challenge for curriculum developers to present to students. Evolution can occur on a small scale, as in microevolutionary events, or on a large, macroevolutionary, scale. Whereas mutations, gene flow, genetic drift, and nonrandom mating may affect the frequencies of alleles in a population, natural selection is the major driving force behind evolution.

The following questions are a few of the major **contemporary issues in the patterns and products of change.** These questions represent active areas of research. For instance, what is the fate of rare and endangered species? Should they be protected and, if so, how and by whom? What is their value, if any, and how is their value determined?

What is the rate of evolution? Does evolution proceed at a slow, steady rate in a process where small changes accumulate to produce larger ones as suggested by the term *gradualism*? Is evolution characterized by long periods of very little change in a population interrupted by a short period of rapid change as suggested by the term *punctuated equilibria*? Or does evolution happen both ways? Scientists agree that evolution occurs, but the rate at which it occurs still is being debated.

Will the Earth experience another mass extinction of its living organisms? Paleontologists have identified several major mass extinctions such as the one that eliminated the dinosaurs approximately 65 million years ago. Do these mass extinctions occur on a regular basis, and, if so, what are their causes? Does extinction result from "bad genes, or bad luck"?

What will be the role of genetic engineering in the future with regard to the development of new plants and animals—crop plants and domesticated animals—and to the development of medicines for humans?

Did the human races arise from Africa and spread to other parts of the world or did the races arise in different parts of the world, each race adapting to a different set of environmental conditions?

How many species of plants and animals remain unidentified? Will they be identified and studied before the environment in which they evolved is destroyed, and which of them are potentially useful to humans as sources of food or medicine?

EVOLUTION: PATTERNS OF CHANGE *Living Systems Change Through Time*	High School	2-Year College	4-Year College
The Dynamic Earth	✓	✓	✓✓
Scales of Time	✓✓	✓✓	✓✓
Forces of Evolutionary Change	✓✓✓	✓✓✓	✓✓✓
Genetic Variation	✓✓✓	✓✓✓	✓✓✓
Natural Selection	✓✓✓	✓✓✓	✓✓✓
Gene Flow/Genetic Drift	✓✓	✓✓	✓✓
Nonrandom Mating	✓	✓	✓
Artificial Selection	✓✓	✓✓	✓✓
• Domesticated plants, animals, & microbes	✓	✓	✓✓
• Advances in biotechnology	✓✓	✓✓	✓✓
Patterns of Evolution	✓✓✓	✓✓✓	✓✓✓
Historical background	✓	✓✓	✓✓
• Cultural vs. biological evolution	✓✓	✓✓	✓✓
Microevolution	✓	✓✓	✓✓
Macroevolution	✓	✓✓	✓✓
• Gradualism/punctuated equilibria	✓	✓	✓
Extinction	✓✓✓	✓✓✓	✓✓✓
Conservation Biology	✓✓✓	✓✓✓	✓✓✓
Renewable & Nonrenewable Resources	✓✓✓	✓✓✓	✓✓✓
Population Genetics	✓✓	✓✓	✓✓
Gene Pool/Gene Frequency	✓✓	✓✓	✓✓
Hardy-Weinberg Equilibrium	✓	✓✓	✓✓
✓✓✓ = ESSENTIAL ✓✓ = OPTIONAL ✓ = NONESSENTIAL			

Table 5-2a: Unifying Principle 1 with related major concepts

EVOLUTION: PRODUCTS OF CHANGE Evolution Has Produced a Diversity of Living Systems on Earth			
	High School	2-Year College	4-Year College
Origin of Life	✓✓	✓✓✓	✓✓✓
Characteristics of Living Systems	✓✓✓	✓✓✓	✓✓✓
Fossils	✓	✓	✓
Specialization & Adaptation	✓✓✓	✓✓✓	✓✓✓
Species & Speciation	✓✓	✓✓	✓✓
Isolating Mechanisms	✓	✓✓	✓✓
Adaptive Radiation	✓✓	✓✓	✓✓
Human Evolution	✓✓✓	✓✓✓	✓✓
• Diversity of people & cultures	✓✓	✓✓	✓✓
Phylogenetic Classification	✓✓	✓✓✓	✓✓✓
Homologous & Analogous Structures	✓✓	✓✓	✓✓
Biodiversity	✓✓✓	✓✓✓	✓✓✓
Prokaryotes	✓✓	✓✓	✓✓
Eukaryotes	✓✓	✓✓	✓✓
• Protists	✓	✓✓	✓✓
• Fungi	✓	✓✓	✓✓
• Plants	✓✓	✓✓	✓✓
• Animals	✓✓	✓✓	✓✓
Biogeography	✓	✓✓	✓✓
✓✓✓ = ESSENTIAL ✓✓ = OPTIONAL ✓ = NONESSENTIAL			

Table 5-2b: Unifying Principle 1 with related major concepts

PRINCIPLE 2. *Interaction and Interdependence*
Living Systems Interact with Their Environment and Are Interdependent with Other Systems

Organisms must obtain the necessary resources of life from their environment, which consists of biotic and abiotic factors. Ecology is the study of the relationship between an organism and these factors—how the organism affects and is affected by the world in which it lives. The ecology of an organism includes the methods and mechanisms it uses to obtain nutrients and energy and the interactions with other organisms that compete with it, eat it, or break down its body after it has died. Both biotic and abiotic factors limit the ability of an organism to live and reproduce in a particular ecosystem. By its very existence, an organism exploits its environment and thus modifies it, often to its disadvantage. An organism, however, also may thrive in a particular physical setting and may benefit from its interaction with other species.

Population ecologists study the growth or regulation of a group of organisms of the same species and the life history of that species. Community ecologists study the interactions among species. Because all species of a community are linked to others by their dependence on resources, community ecologists explore such interactions as predation, parasitism, or mutualism. The links established among organisms on the basis of their requirement for energy and nutrients describe the web of life. Another aspect of the community is the change, or succession, in species composition through time. Finally, the study of the biosphere and the physical environment includes large-scale patterns of plant associations within biomes, distinctive major communities found around the world.

The following questions are a few of the **contemporary issues in interaction and interdependence.** For instance, what is the relative contribution of the environment and of genes to alcoholism, aging, and obesity? What are the effects of environmental pollutants such as cigarette smoke on the health of smokers, and on nonsmokers who inhale passive smoke? Is there any danger to human health from environmental concerns such as nuclear power plants and electromagnetic radiation?

What is the role of natural disturbances, such as a fire, on the ecology of a community? How do human disturbances such as the clearcutting of a forest alter an ecosystem, and what impact do these disturbances have on organisms in the ecosystem? Are there any benefits to such disturbances, and what role do they play in the normal function and succession of a community?

How can microorganisms that occur naturally in the soil help reduce the amount of waste in our landfills and of environmental pollutants such as oil spills? What limits the suitability and effectiveness of using microbes in different environmental conditions?

How do pathogens cause disease? What is the importance of a pathogen's genetic makeup, growth, virulence, motility, and resistance to the host's defense mechanisms in relation to its ability to enter a host and adhere to its tissues and cells? What is the interaction between the invading pathogen and the host's immune system that eventually leads to disease?

How are plants able to withstand adverse environmental conditions such as drought and freezing temperatures? What are the structural, molecular, and physiological bases that provide plants with the ability to cope with environmental stresses? Will an understanding of these bases enable us to improve crops?

To what degree and in what ways do species within a community interact? Is a community simply an assemblage of plants, animals, and microorganisms, or is it a complex system made up of many interconnecting networks? What significance might any interconnections have on the community as it changes through time—for example, in succession?

How can microecosystems be used as models to study larger, natural systems and ecological interactions? How can scientists develop an accurate picture of the multiple interactions of natural communities that may have thousands of different species of plants and animals?

In the essay that follows, Eugene Odum addresses some of these questions as he outlines twenty major ideas of ecology that deal with the interaction and interdependence of living systems.

ESSAY

Ecology in the 1990s

by Eugene P. Odum

In the 40 years since the publication of the first edition of my *Fundamentals of Ecology*, the field of ecology has emerged from its roots in biology to become a discipline that integrates organisms, the physical environment, and humans. This emergence is in line with the Greek root of the word "ecology," which is *oikos*, meaning "household." Accordingly, ecology translates as the study of our "house," or biosphere, in which we live and in which we build our human-made structures and cultivate our crops. Because the life-supporting functions of our house are endangered by pollution, wasteful use of resources, and human population pressure, we need an expanded knowledge of ecology to help us become better stewards, i.e., "keepers of the house."

The emergence of ecology as a more global discipline means that the ecosystem level of organization becomes the major focus, with populations and individuals considered as components. That this viewpoint is now generally accepted is indicated by a recent British Ecological Society survey in which members were asked what they considered the most important ecological concepts. "Ecosystem" was the concept most frequently listed.

Accordingly, environmental education, beginning in grade school, should go beyond consideration of species,

biodiversity, food chains, and the litany of environmental problems (acid rain, ozone hole, toxic wastes, etc.) to include the common denominator functions such as energetics, cycling, growth and development, competition and mutualism, and so on. Especially important is the interface between ecology and economics because it is essential that everyone understands the interdependence between the market human-made goods and services and the nonmarket goods and services of nature that provide basic life support (breathing, drinking, and eating).

What we have learned from ecological studies during the past several decades can help us avoid the "quick-fix" and deal with human predicaments on a more long-term basis. For example, how to prosper in a world of limited resources; when to confront and when to cooperate; when to shift from quantitative to qualitative growth; and how to be a prudent parasite that does not destroy its host, the Earth.

The following are twenty concepts, "great ideas" in ecology, for the 1990s and beyond. They might be considered as appropriate knowledge for an environmentally literate individual.

1. *Ecosystems*. An ecosystem is a thermodynamically open system, far from equilibrium. In considering a forest tract, for example, what is *coming in and going out*, especially energy,

materials, and organisms, is equally as important as what is *inside* the tract. The same holds for a city; it is not a self-contained unit ecologically or economically. Its future depends as much on the external life-support environment as on activities within city limits.

2. *Source-Sink Concept*. This is a corollary to Concept 1 and is applicable at the ecosystem as well as the population level. At the former level, an area of high productivity (salt marsh, for example) may "feed" an area of low productivity (adjacent coastal waters). At the population level, a species in one area may have a higher reproduction rate than needed to sustain the population, and surplus individuals then may provide recruitment for an adjacent area of low reproduction. Food chains also may involve sources and sinks; see Concept 12.

3. *Hierarchical Organization of Ecosystems*. Species interactions, such as interspecific competition, herbivore-plant, and predator-prey, that tend to be unstable are constrained by the slower interactions that characterize large systems, such as oceans, the atmosphere, soils, and large forests. Large, complex systems tend to go from randomness to order. Accordingly, ecosystems as a whole tend to be more homeostatic than their components. This may be the most important principle of all because it warns that what is true at one level may not be true at another level of organization.

4. *Cybernetics at the Ecosystem Level*. At the ecosystem level, cybernetics differs from that at the organism level or the level of human-made mechanical systems (temperature control of a building, for example) in that feedback is internal and not goal oriented rather than external with a set point. This is a corollary of Concept 3.

5. *Natural Selection at More Than One Level*. Natural selection may occur at more than one level, another corollary to Concept 3. Accordingly, co-evolution, group selection, and traditional Darwinism all are part of the hierarchical theory of evolution.

6. *Types of Natural Selection*. There are two types of natural selection: (a) organism vs organism, which leads to competition; and (b) organism vs environment, which leads to mutualism.

7. *Competition*. Competition plays a major role in shaping the species composition of biotic communities, but competition exclusion in the open systems of nature is probably the exception rather than the rule because species often are able to shift their functional niches to avoid the deleterious effects of competition.

8. *Evolution of Mutualism*. Mutualism increases when resources become scarce either because they are tied up in the biomass, as in mature forests, or because soil or water is nutrient poor, as in some coral reefs or rain forests. This concept is a follow-up of Concept 6 and has relevance in connection with the shift from confrontation to cooperation among superpowers.

9. *Network Mutualism Concept*. When a predator-prey food chain functions in a food web network, organisms at each end (for example, plankton and bass in a pond) are indirectly mutualistic in that they benefit each other. Accordingly, there are both positive and negative interactions in a food web network.

10. *Stress Theory*. First signs of stress (early warning) usually occur at the population level but consequences become serious when ecosystem-level effects can be detected. This is another corollary of Concept 3, namely, that parts are less stable than wholes.

11. *Modified Gaia Hypothesis*. This hypothesis, that organisms since early in geological history have interacted in a positive way with the physical environment to make the Earth more livable by increasing O_2, and decreasing CO_2 and pH, is now accepted by many scientists. Especially accepted is the concept that microorganisms play major roles in vital cycles (e.g., the nitrogen cycle) and

in atmospheric and oceanic homeostasis.

12. *Bacterial Control*. Bacteria control the energy flow in marine food webs. This is a corollary to Concept 2. In warm waters, bacteria function as a "sink" in that they may short-circuit the energy flow so that less energy reaches the bottom to support fisheries.

13. *Expanded Approach to Biodiversity*. Biodiversity includes genetic and landscape diversity, not just species diversity.

14. *Ecosystem Development*. Ecosystem development, or autogenic ecological succession, is a two-phased process. Early or pioneer stages tend to be stochastic as opportunistic species colonize, but later stages tend to be more self-organized (perhaps another corollary of Concept 3).

15. *Carrying Capacity*. Carrying capacity is a two-dimensional concept involving number of users (density) and intensity of per capita use that track in a reciprocal manner. As the intensity of per capita impact goes up, the number of individuals that can be supported by a given resource base goes down. Recognition of this principle is important in determining how much of a natural environment buffer to set aside in land-use planning.

16. *Input Management Concept*. In the past, increasing the yield of food, electricity, and other outputs from production systems (e.g., crop fields, power plants) has been accomplished by increasing the inputs of energy and chemicals which, in turn, has increased the output of nonpoint pollution and other waste. The only way to deal with this situation is to reduce the input of environmentally damaging materials by input management. In other words, waste reduction rather than waste disposal must now be the focus of technology.

17. *Net Energy Concept*. Expenditure of energy (the energy penalty) is always required to produce or maintain an energy flow or a material cycle. As communities and systems become larger and more complex, then more and more of the available energy is required for maintenance (respiration). This is a fact that city managers often fail to recognize when it comes to high energy, complex urban-industrial developments.

18. *Goods and Services*. There is an urgent need to bridge the gap between market (human-made) and nonmarket (natural life-support) goods and services; and between nonsustainable short-term and sustainable long-term management (agroecosystems and tropical forests of special concern). H. T. Odum's concept of Emergy (spelled with an *m*) and Daly-Cobb's Index of Sustainable Economic Welfare are examples of recent attempts to bridge these gaps.

19. *Transition Costs*. These costs always are associated with major changes in nature and in human affairs. Society has to decide who pays for such expenditures as the cost in new equipment, procedures, and education in changing from high-input to low-input farming or converting from air-polluting to clean power plants.

20. *A Parasite-Host Model for Man and the Biosphere*. This is the basis for turning from dominionship (exploiting the Earth) to stewardship (taking care of the Earth). Despite technological achievements, humans remain parasitic on the biosphere for life support. Survival of a parasite depends on (a) reducing virulence and (b) establishing reward feedback that benefits the host. Similar relationships hold for plant-herbivore or producer-consumer interactions. In terms of human affairs, this means reducing waste of resources and promoting the sustainability of renewable ones.

INTERACTION & INTERDEPENDENCE *Living Systems Interact with Their Environment & Are* *Interdependent with Other Systems*	High School	2-Year College	4-Year College
Environmental Factors	✓✓✓	✓✓✓	✓✓✓
Biotic & Abiotic Environmental Factors	✓✓	✓✓✓	✓✓✓
Adaptation	✓✓✓	✓✓✓	✓✓✓
Limiting Factors	✓✓✓	✓✓✓	✓✓✓
Population Ecology	✓✓	✓✓	✓✓
Population Attributes	✓✓	✓✓	✓✓
Population Regulation Factors	✓✓	✓✓	✓✓
Carrying Capacity	✓✓✓	✓✓✓	✓✓✓
Plant & Animal Populations	✓	✓	✓✓
Community Structure	✓✓✓	✓✓✓	✓✓✓
Species Richness & Species Diversity	✓	✓	✓✓
Food Chains & Food Webs	✓✓✓	✓✓✓	✓✓✓
• Producers, consumers, decomposers	✓✓✓	✓✓✓	✓✓✓
Niche	✓✓	✓✓	✓✓
Interactions Among Living Systems	✓✓✓	✓✓✓	✓✓✓
• Predation/parasitism	✓	✓	✓✓
• Mutualism/coevolution	✓	✓	✓✓
• Competition	✓	✓	✓✓
Ecosystems	✓✓✓	✓✓✓	✓✓✓
Nutrient & Water Cycles	✓✓✓	✓✓✓	✓✓✓
Energy Flow	✓✓✓	✓✓✓	✓✓✓
Biomes	✓✓	✓✓	✓✓
Succession	✓✓	✓✓	✓✓
Biosphere	✓✓✓	✓✓✓	✓✓✓
Human Influences on the Biosphere	✓✓✓	✓✓✓	✓✓✓
• Human-designed ecosystems	✓✓✓	✓✓✓	✓✓✓
− Agriculture & food production	✓✓✓	✓✓✓	✓✓✓
• Overpopulation	✓✓✓	✓✓✓	✓✓✓
• Resource use & production of waste	✓✓✓	✓✓✓	✓✓✓
• Atmosphere & habitat alteration	✓✓✓	✓✓✓	✓✓✓
✓✓✓ = ESSENTIAL ✓✓ = OPTIONAL ✓ = NONESSENTIAL			

Table 5-3: Unifying Principle 2 with related major concepts

PRINCIPLE 3. *Genetic Continuity and Reproduction*
Living Systems Are Related to Other Generations by Genetic Material Passed on Through Reproduction

Life is a continuing stream of genetic information passed from generation to generation. Through reproduction, organisms produce other, similar organisms, maintaining the line of genetic continuity linking generations. Reproduction may involve one parent (usually asexual reproduction) or two parents (sexual reproduction). In eukaryotes, sexual reproduction begins with the production of special reproductive cells, or gametes, through a process called meiosis and, later, the fusing of two gametes in fertilization to form a zygote. The type of reproduction depends on the species of organism, although some organisms may switch from one form to another depending on environmental factors.

Gametes contain the information for the development of characteristics passed from one generation to the next. Geneticists can determine the parents' genetic makeup by studying the patterns of inheritance revealed in the characteristics of the offspring. The biochemical basis of heredity is a set of genes made of the nucleic acid DNA and located on chromosomes. DNA is an information molecule. Each gene is a unique coded message formed by a combination of four types of subunits termed nucleotides. A gene directs the production of a specific polypeptide that regulates other genes, catalyzes a reaction leading to a distinct characteristic of an organism, or constitutes a needed structural component of the cell. Gene regulation involves the gene's expression and control, and this regulation is influenced by the environment of the cell in which the gene is located. Although different organisms have different individual coded messages, the basis for all genetic codes is virtually universal. That is, the same types of nucleotides make up DNA in all organisms, and the methods of transcription and translation from code to characteristic are similar in nearly all living organisms.

DNA is stable, generally replicating with great fidelity. The four bases of DNA, in their multiple combinations, harbor all of the genetic variation represented by life on Earth. Genetic control mechanisms provide for duplication of living forms, but they also may introduce variations through mutations and genetic recombinations that create new material for natural selection, leading to evolution. Modern genetic engineering technology has allowed humans to rearrange genetic material in an organism and to move genetic material from one organism to another, creating new combinations of genes. It also has allowed scientists to study the genetic material in minute detail.

The following questions are a few of the **contemporary issues in genetic continuity and reproduction**. For instance, what is the molecular structure of the gene, and how does it replicate? How are genes organized on chromosomes, and of what importance is this organization to the normal functioning of the gene? What is the function of so-called "junk" DNA? What turns a gene on and off?

How do genes control an organism's development and behavior? How are scientists able to identify genes related to developmental defects in humans?

What is the importance of preserving the wild relatives of domesticated plants and animals? Can particular genes from these wild organisms be transferred into crop plants or animals that would then confer resistance to disease or increased productivity to the domesticated organism? Or will this genetic diversity be lost with the loss of the wild plants and animals?

What significance will gene therapy play in the future treatment of human genetic diseases? Can the growth of cancer cells be inhibited by inserting normal genes into the cells of cancer patients who lack such genetic material?

How do genes normally found in human cells and with life-sustaining regulatory functions become oncogenes—cancer-causing agents? How can the immune system of humans be regulated and stimulated to help fight off cancers and other diseases? What is the role of genes in the leading causes of death and disability in developed countries, for example, heart disease, stroke, cancer, and diabetes? What is the role of genes in susceptibility to infectious disease? What role do genes play in complex human traits such as intelligence, creativity, or aggression?

What advances in biotechnology will come as our understanding increases about genetics and the biology of living organisms? What new technologies such as protoplast fusion, tissue culture, somaclonal variation, and genetic engineering may help develop useful organisms with unique and novel combinations of genes?

The following essay by Francisco Ayala discusses many of the aspects of genetics discussed in this section as well as the bases of genetic continuity for all living organisms.

ESSAY

Genetic Continuity

by Francisco J. Ayala

People have known from antiquity that children resemble their parents. But the laws of biological heredity were only discovered in the middle of the nineteenth century by Gregor Mendel, an Augustinian monk and teacher who lived in Brünn, Austria (now Brno, Czechoslovakia). Mendel used garden peas (*Pisum sativum*) in his experiments, but his discoveries apply to other plants and animals as well.

Mendel discovered that each trait's inheritance is determined by the transmission of two factors (later called *genes*), one inherited from each parent. For example, the color of a pea is determined by the genes for pea color, which can be yellow or green. When the two genes for a trait are identical (say, yellow-determining genes), the trait exhibits that particular configuration (that is, the pea is yellow). When the two genes are different, only a particular one of the two is expressed (thus, a pea that carries one gene for yellow and one for green is, in fact, yellow). The gene that is expressed is called *dominant*, the one that remains concealed is *recessive* (so the gene for yellow is dominant over the gene for green). The genetic constitution of an individual is called *genotype*; its physical appearance is the *phenotype*.

Mendel discovered that the two genes for a trait separate in the sex cells (called *gametes*) so that only one of the two genes is present in each gamete. Thus, a gamete can carry the gene for yellow or the gene for green, but not both. This discovery is known as the "Law of Segregation." When fertilization occurs, one gene coming from the mother joins one coming from the father so that the new individual carries two genes for each trait, just as the parents did. If one parent passes on the gene for yellow and the other the gene for green, the progeny pea is yellow (because of the dominance of yellow over green); if both parents pass on the gene for green, the pea is green; and if both parents pass on the gene for yellow, the pea is yellow.

Mendel also discovered that genes for different traits (such as the genes for pea color and pea shape) are inherited independently from each other. This discovery is known as the "Law of Independent Assortment." The important implication of this discovery is that in the gametes, the genes inherited from one parent become reassorted with those inherited from the other parent, so that each gamete carries a different random combination of the genes inherited from the two parents. Consider a plant that has inherited the genes for yellow color and round shape from one parent and those for green and wrinkled from the other parent; the gametes produced by this plant are yellow-round, yellow-wrinkled, green-round,

and green-wrinkled in equal proportions. For two pairs of genes the number of kinds of gametes is four (2^2), for n pairs of genes the number is 2^n, which becomes an astronomical number when n is in the thousands as it is in organisms. The Law of Independent Assortment does not always strictly apply, because genes proximally located to each other on a chromosome (chromosomes are the cell components carrying the genes) are more likely to be transmitted together than those located far apart or on different chromosomes. In any case, a sexually reproducing organism produces virtually infinite kinds of gametes. Hence, all individuals are genetically different from one another because they are formed by the fusion of two gametes that are always different from any two other gametes. Biological individuality is an important property of humans and other sexual organisms that derives from the laws of biological inheritance.

Not all organisms reproduce by intercrossing. Some reproduce by self-fertilization, such as some plants in which the sperm and egg come from the same individual. Single-cell organisms such as bacteria and many protozoa reproduce by clonal division: the genes double and the cell divides into two, in such a way that each daughter cell receives one of the two identical sets of genes. In clonal organisms, parents and progeny are all genetically identical to one another (just like all the cells in the body of a multicellular plant or animal), except for an occasional "mutation."

Mutation is the process by which new genetic variants arise. There may be single-gene, as well as chromosomal, mutations. An example of a single-gene mutation is the change of the gene coding for yellow into the gene coding for green peas. An example of chromosomal mutation is found in some people who carry a fraction of chromosome 21 attached to chromosome 14. Mutations are often harmful: an example is the mutation that changes the gene for normal hemoglobin into the gene for sickle-cell anemia. But many mutations are neutral and some are beneficial. Whether or not a mutation is beneficial may depend on the environment. An example is the mutation changing the color of moths in England from "peppered" to black, which is beneficial in soot-covered industrial regions, but not in unpolluted areas.

It was discovered in the mid-twentieth century that genes are molecules of deoxyribonucleic acid (DNA), a chemical consisting of four kinds of units called *nucleotides*. Each nucleotide is made up of three components: phosphoric acid, a simple sugar (deoxyribose), and a nitrogen base. The only difference among the four nucleotides is the nitrogen base, which can be one of four: adenine (A), cytosine (C), guanine (G), and thymine (T). The DNA is organized as a long "double helix" consisting of two complementary strands. The genetic information is encoded in the sequence of nucleotides (e.g., ATTCG...) very much in the same way as semantic information is encoded in the sequence of letters of an English text. A gene mutation occurs when one or a few nucleotides in the DNA sequence are changed.

The DNA sequence of a gene is thousands of nucleotides long. The number of possible different sequences n-nucleotides long is 4^n, which is an astronomical number when n is in the thousands. Thus, there is no practical limit to the number of different "messages" that can be encoded in the DNA of a gene. The DNA in a human cell is several billion nucleotides in length; in many animals and plants it ranges from about one hundred million to several billions. The number of gene pairs encoded by the DNA is estimated at about one hundred thousand for humans and from ten to one hundred thousand in most animals and plants.

The two strands of the DNA double helix are complementary. Hydrogen bonds link each nitrogen base in one strand with a nitrogen base in the complementary strand in such a way that A always pairs with T, and C with G. Rep-

lication of the DNA occurs by separation of the two strands and synthesis of two new strands, which, by the pairing rules, are identical to the parental strands. For example, if the sequence in one strand is ATTCG..., the complementary strand is TAAGC... When these two strands separate, the first one acts as a template for the synthesis of a new strand that is precisely like the second strand. The second strand directs the synthesis of a strand that is precisely like the first one. Thus, the two daughter DNA molecules are identical to each other and to the parental one. The pairing rules between the nitrogen bases account for the fidelity of the process of reproduction. Variation is introduced by the processes of mutation and recombination. The spread and accumulation of favorable mutations account for the evolution of organisms.

The applications of genetics to human life are multifarious. Animal and plant breeders produce breeds and crops with desired properties, such as rapidly growing chickens, protein-rich corn, and seedless grapes. Genetic knowledge is necessary for characterization, diagnosis, and treatment of the thousands of human diseases that derive from genetic defects, such as hemophilia and cystic fibrosis. Genetics furnishes the scientific basis for understanding human diversity, the origin of species, and the process of evolution.

The newly discovered techniques of "recombinant DNA" have opened up possibilities not even dreamed a few years ago. Bacteria have been "engineered" that carry the human gene for insulin and synthesize the insulin needed for treating diabetics. Other bacteria have been genetically designed for digesting oil leaks and spills. Defective human genes can be replaced by functional ones in abnormal cells, so that the bone marrow of a sickle-cell patient will produce normal hemoglobin. The applications of genetic engineering extend beyond the living world to industrial processes and virtually every human activity. Recombinant

DNA and genetic engineering are very recent discoveries. We barely can begin to discern the multitude of possible beneficial applications.

The Human Genome Project is a program in the United States for mapping the position of all human genes onto their chromosomes, and for ascertaining the DNA sequence of the three billion or so nucleotides in the human genome. This arduous and costly enterprise will facilitate understanding of how genes are organized and function and, very importantly, will be extremely beneficial for human health because it will lead to the identification of the genetic causes of disease.

GENETIC CONTINUITY & REPRODUCTION *Living Systems Are Related to Other Generations by Genetic Material Passed on Through Reproduction*	High School	2-Year College	4-Year College
The Gene	✓✓✓	✓✓✓	✓✓✓
Historical Development of the Concept	✓	✓	✓
Molecular Structure	✓✓	✓✓	✓✓
DNA (The Genetic Material)	✓✓✓	✓✓✓	✓✓✓
Replication	✓✓	✓✓	✓✓
Mutations & Mutagens	✓✓	✓✓✓	✓✓✓
Gene Action	✓✓	✓✓✓	✓✓✓
Transcription	✓✓	✓✓	✓✓
RNA	✓✓	✓✓	✓✓✓
Translation	✓✓	✓✓	✓✓
Gene Regulation	✓✓	✓✓	✓✓✓
Interaction of Genotype & Environment	✓✓✓	✓✓✓	✓✓✓
Patterns of Inheritance	✓✓✓	✓✓✓	✓✓✓
Mendelian Genetics	✓✓	✓✓	✓✓✓
• Dominance	✓✓	✓✓	✓✓
• Independent Assortment/Recombination	✓✓	✓✓	✓✓
NonMendelian Genetics	✓	✓✓	✓✓
Human Genetics	✓✓	✓✓	✓✓
Reproduction	✓✓✓	✓✓✓	✓✓✓
Asexual Reproduction	✓✓	✓✓	✓✓
• Mitosis	✓✓	✓✓✓	✓✓✓
✓✓✓ = ESSENTIAL ✓✓ = OPTIONAL ✓ = NONESSENTIAL			

Table 5-4: Unifying Principle 3 with related major concepts

GENETIC CONTINUITY & REPRODUCTION Living Systems Are Related to Other Generations by Genetic Material Passed on Through Reproduction			
	High School	2-Year College	4-Year College
Sexual Reproduction	✓✓✓	✓✓✓	✓✓✓
• Meiosis/Fertilization	✓✓	✓✓✓	✓✓✓
Reproductive Systems	✓✓	✓✓	✓✓
• Morphological	✓	✓✓	✓✓
• Physiological	✓	✓✓	✓✓
• Human Reproduction	✓✓✓	✓✓✓	✓✓
— Reproductive cycle	✓✓✓	✓✓✓	✓✓
— Contraception	✓✓✓	✓✓✓	✓✓
— Prenatal development	✓✓✓	✓✓✓	✓✓
— Reproduction & biotechnology	✓✓✓	✓✓✓	✓✓
Molecular Genetics	✓	✓✓	✓✓
Microbial/Viral	✓	✓	✓✓
Eukaryotic	✓	✓	✓
Genetics & Biotechnology	✓✓✓	✓✓✓	✓✓✓
Genetic Engineering	✓✓✓	✓✓✓	✓✓✓
Human Genome Project	✓✓	✓	✓
Gene Resources & Gene Banks	✓✓	✓✓	✓✓
✓✓✓ = ESSENTIAL ✓✓ = OPTIONAL ✓ = NONESSENTIAL			

Table 5-4 Continued

PRINCIPLE 4. *Growth, Development, and Differentiation*
Living Systems Grow, Develop, and Differentiate During Their Lifetimes Based on a Genetic Plan That Is Influenced by the Environment

Most multicellular organisms begin as a zygote, the single cell produced at fertilization that grows into a mature organism. The zygote first develops into a multicellular embryo through mitosis, the cell division that increases the number of body cells, followed by an increase in cell size. Thus, the zygote must contain all the genetic information necessary to form a complete organism, and this information must be passed from cell to cell in the developing body. In all multicellular organisms, cells begin to differentiate through time, taking on specialized roles. Division of labor at the cellular level begins during the early development of an organism and continues throughout its life. Eventually, morphogenesis—the growth and development of body parts—takes place so that the adult has many different structures with unique functions. In his essay, John Tyler Bonner describes the important questions of contemporary developmental biology and various factors that contribute to the progression of a single cell into a multicellular adult organism.

Growth does not, however, follow a haphazard course but conforms to a well-defined pattern of differentiation controlled by an organism's particular genetic makeup. Animal development proceeds through an orderly series of stages after fertilization, including the embryonic stages of the blastula and the gastrula. Plants have an open-ended, indeterminate growth pattern. That is, at least part of the plant's body continues to grow and develop throughout its life and, although the development of the plant body is under genetic control, the environment can have a major influence in the final shape the plant takes. Because no organism lives isolated from the physical world, its genetic plan for growth and development and the rate at which this plan unfolds also are affected by the environment. Growth and development are powered by an intake of energy and are supported by the uptake of material resources. Thus, the body of an adult organism is characterized by structures that optimize the intake, transport, and utilization of necessary resources.

Organisms that obtain an ample share of resources have the potential to produce a greater number of offspring than their less successful competitors. One key to such a successful harvest of resources is efficiency. Division of labor at the level of cells can lead to such efficiency, a quality found in both unicellular and multicellular organisms. When an organ, a cell, or a molecule assumes a distinct role, the function of the structure may be greatly enhanced. The efficiency of a structure can be optimized by its form: the broadness of leaves can contribute to more efficient capture of sunlight; the shape of molars helps to grind food. Natural selection favors individuals with structures that are better suited to carrying out the demands of daily living, so the morphological characteristics of an organism contribute to its physiological needs. Not all organisms, however, have achieved the same resolution to environmental problems, so there are numerous adaptations for obtaining and using resources, many of them morphological. The environment may even directly affect the morphology of an organism, with limits to such changes in form imposed by the individual's genetic constitution. Whereas the form may help reveal function, it also is true that the function and habitat to which an organism has adapted can be inferred from the anatomy or morphology of a given structure. The enhancement of function due to the form of a structure is a phenomenon that is seen at the molecular, cellular, tissue, and organ level in the hierarchy of organization as well as in the overall morphology of the individual.

The following questions are among many **contemporary issues in growth, development, and differentiation.** For instance, what are the uses of human fetal tissue in curing human diseases, and what are the biological, social, and bioethical issues associated with the use of this tissue?

What controls the developmental pattern in multicellular organisms? Do all organisms possess "master control genes" that direct and regulate their developmental process, controlling their entire body plan? How do these control genes affect the expression of other genes?

How is senescence, the genetically controlled sequence of biochemical events leading to death, controlled in living organisms? Can senescence be delayed so the productivity of a crop plant increases and the shelf-life of its fruits or vegetables lengthens? What is the interaction of hormones, receptors, and enzymes in the senescence of a plant?

What determines the fate of a cell in a developing embryo? A developing cell may differentiate into a brain cell or a liver cell, but what determines this differentiation? It is known that developmental control of gene expression within such cells depends on the time in the developmental process and position of the cell in relation to others, but how important are cell lineage (the developmental history of a cell from the zygote stage) and the immediate cell environment?

What determines the life span of individual cells and organisms? Aging and death are normal biological processes that are under at least some genetic control, but how much? Can these processes be reversed?

How is the structure of macromolecules, such as complex proteins, important to their function? What scientific questions might be answered by the sequencing of molecules of known function, such as determining the exact sequence of amino acids in a particular protein?

Other important questions about growth, development, and differentiation are identified in Christian de Duve's essay that addresses a central problem of biology—to explain function by form and, reciprocally, to convert description of form into understanding of function.

ESSAY

Development

by John Tyler Bonner

Development is a general term used to describe the period of the life cycle of any organism as it progresses from a single cell stage (either a fertilized egg or an asexual spore) to a multicellular adult organism. The study of this period of size increase used to be called embryology, but because all organisms, not just higher animals and plants, undergo this kind of change, development is a more useful, all-encompassing term that can include protozoa, algae, fungi, and other lower forms of life.

In every multicellular life cycle there is a remarkable transformation of a single cell to a complex organism. Clearly this is a fascinating question: where do the instructions for development come from in the initial single cell, and what are the rules by which they are carried out? It is a problem that has intrigued biologists including Aristotle (and no doubt others before him) as well as many of us today. Our views on development have gone through many permutations, but modern developmental biology started in the middle of the last century, and has undergone a second revolution, built on the first, in the last ten to twenty years. Let me describe these two revolutions and their relation to one another, and then discuss why these problems seem so important to us right now and how they fit in with the rest of biology.

The last century was a period of enormously important discoveries for developmental biology. It began with the realization that all living things were made up of cells, and later, that all those cells contained chromosomes in their nucleus that were duplicated and passed on to the daughter cells after each division. The significance of these findings was not fully appreciated until the beginning of this century when Mendel's laws of heredity were suddenly "rediscovered" by the whole world, and it was clear that the hereditary units of Mendel, the genes, were on the chromosomes of the cells. Meanwhile, experimental developmental biology was born. Biologists demonstrated that in some developing eggs there was a rigid, fixed pattern of cell cleavage, and that some of the resulting cells developed into specific structures in the embryo, such as muscle or nerve tissue. If those cells are removed at an early stage, the resulting embryo (which does not live to maturity) will lack muscle or nerve, or the descendants of whatever cells had been removed. This is called *mosaic* development. While this work from the American school of embryology was in progress, a European worker was finding something quite different: the embryos of some animals could be cut in two at an early stage of development and form two perfectly proportioned dwarf embryos (and not half embryos). In this case the fates of cells and their descendants did not seem to be fixed, but could change in a way that showed they were aware of their position within the whole embryo and differentiated into muscle and nerve, and so forth, because they were in the right place. This is called *regulative* development.

At first there seemed to be an absolute division line between mosaic and regulative development. This line, however, has eroded for three reasons. First, biologists found that even the most mosaic embryos showed some regulation in their development. Secondly, some organisms appeared to be mosaic at early stages of their development, but later they became wholly regulative, or vice-versa. The third reason is that in this century, scientists have progressively appreciated that in some organisms the cells in the embryo will take on specific characteristics (that is, they are essentially mosaic cells), but they wander about in the embryo and sort themselves out. So in some embryos (especially those of vertebrates) there are coordinated movements and adhesions between cells that result in the migrating of cells of different characters to specific places to produce a pattern.

Perhaps the most important discovery of the old embryology in the early part of this century was that of *induction*: the demonstration that one part of the embryo could send signals to another and persuade it to develop into specific cell types and specific structures. This discovery won a Nobel prize for Hans Spemann. This was the beginning of an active field of inquiry that is a central part of developmental biology today—what are the inducing chemicals (now often called *morphogens*), and how do they work?

All of which brings us to the *new* developmental biology that can be traced to the 1950s and the discovery by James Watson and Francis Crick of the structure of DNA, the structure of genes. Their work has led to one of the most exciting and fast-moving fields, that of molecular developmental biology. In fact there has been an extraordinary merger of molecular biology, cell biology, and genetics, all to focus on the big questions of developmental biology. One begins with the obvious idea that all the instructions for development must be in the genes on the chromosomes in the fertilized egg. (This turns out to be not quite true—the egg cytoplasm also carries some essential instructions, but many of these are derived from maternal genes.) Therefore, the place to look for a plan of development must be in the expression of genes during the course of development. This kind of research is being done with noted success, for example, on the

development of the fruit fly and of a nematode worm where, in both cases, much is known about their genes. We now know the location of numerous genes on the chromosomes, including their DNA structure. We know the structure of the proteins they produce and, due to the wonders of modern techniques, we know where and when in the embryo those proteins appear. The techniques are tremendously powerful and at present the progress of developmental biology is swift and of absorbing interest.

There are, however, two important things to remember and appreciate. One is that with all this recent progress, the question of how one gets a consistent pattern out of a plethora of gene actions has not simplified our picture of the mechanism of development; if anything it has made it more complicated. Here we have been greatly helped, and will no doubt be more so in the future, by applied mathematicians who can make simplifying models that are helpful in giving order and understanding to the great variety of empirical facts that keep streaming in. The models may be naive, and sometimes even incorrect, but at least they show us how, in theory, there are possible straightforward ways in which genes could control development.

The other major point to remember is that there can be many developmental steps that are completely out of the reach of genes. The genes within the egg set up the initial conditions, and those genes appearing later have some additional effects, but most of the action takes place way beyond their influence. This does not mean genes do not control, but to get the desired result, be it a mammal or a tree, we must take into account many steps that will occur far removed from the primary proteins the genes produce. To give an example, genes are responsible for producing the glycoproteins on the surface of cells of animal embryos, and those glycoproteins will affect the relative adhesiveness of the cells. In turn, the degree of adhesiveness of different cell types will affect the sorting of the cells and their ultimate pattern. One cannot build a building with the blueprint alone, and this same principle applies to development. All the properties of bricks, mortar, steel, wood, and other parts of the house are essential for achieving the final "adult" form of the house.

As we can see from this bird's-eye view of modern developmental biology, it brings together genetics, cell biology, molecular biology, biochemistry, and, when one considers specifically the development of the nervous system, neurobiology. Because it expands into so many areas, the study of development is central to all biology. This includes evolution. Development is, after all, an important part of the life cycle.

To say the same thing differently: the principal way to alter adults, and therefore their reproductive success, is to alter their development. In order for Darwinian natural selection to work, changes must be inherited, that is, genetic, and therefore selection works on the genes that control development.

Besides these grand reasons for the importance of development, there is a very practical one as well. Many medical problems are problems of development, cancer being the most obvious, but by no means the only one. If we can obtain a deeper understanding of the basis of development, it will inevitably lead to a better understanding of disease.

ESSAY

Form and Function

by Christian de Duve

Living beings have characteristic shapes and structures, which fall under the heading of *form*. Living beings also accomplish tasks; they perform activities, such as manufacturing specific substances, moving themselves or parts of themselves, and producing offspring. As a rule, these activities are useful to the individual or to the species concerned. They serve a *function*.

A central problem of biology is to explain function by form and, reciprocally, to account for form as a product of function; that is, to convert description into understanding. To do this successfully, scientists have had to follow three rules, at least in a pragmatic, methodological fashion. As I shall point out, scientists remain free to adopt or reject the underlying philosophical doctrines.

The first rule is that living organisms must be considered as machines that can be entirely understood in terms of the laws of physics and chemistry. This approach is called *mechanistic*—or Cartesian, from the name of René Descartes, the 17th century philosopher who first recommended it. Conceptually, mechanism is opposed to vitalism, which posits the intervention of a special vital principle in living beings. Nowadays, very few biologists, if any, adhere to vitalism. Even those who do still have to adopt a mechanistic approach if they wish to engage in a valid scientific investigation. Only by persistently failing in such an approach can one muster support for vitalism. Note that such failure is not on the horizon. On the contrary, the mechanistic approach has proved—and is still proving—spectacularly successful.

The second rule, also laid down by Descartes, is that complex problems, such as those encountered in biology, must be subdivided into as many simpler parts as is possible in order to be tractable. This rule is a logical corollary of the mechanistic approach. The only way to understand the working of some intricate piece of machinery is by pulling it apart and studying each part separately. This *reductionist*, or analytical approach, is criticized by the defenders of holism, who maintain that the whole is more than the sum of its parts and must be studied in its totality. This statement also is applicable to a watch, a motorcar engine, or any other machine. It is true that much can be learned from the observation and investigation of intact living organisms. Disciplines such as ecology, ethology, evolutionary biology, as well as several of the medical sciences, are essentially holistic in their approaches. They must be so to achieve their goals. As such, these disciplines have done descriptive wonders. They even have yielded some profound insights, into the mechanisms of evolution, for example, or into the causes of disease. The fact remains, however, that holistic investigations treat the objects of their studies essentially as black boxes; they do not—and cannot—provide the degree of understanding that one obtains by opening the black box and analyzing its contents by a reductionist approach. The analysis must not stop there, of course. Its findings must be fitted back into a coherent picture explaining the operation and behavior of the black box as a whole. Reduction must be followed by reconstruction, analysis by synthesis. Another famous advocate of the mechanistic-reductionist approach, the 19th century physiologist Claude Bernard, made this abundantly clear in his *Introduction to the Study of Experimental Medicine*.

These considerations lead to the third rule. For complete understanding, explanations correlating form and function must be given in *molecular* terms. The reason for this is simple: living organisms are chemical machines.

Only at the molecular level does structure authentically illuminate function and vice versa.

The latter part of this century has witnessed immense advances in our molecular understanding of life's most basic processes. These acquisitions cannot possibly be ignored in the teaching of biology, even at an elementary level. Fortunately, contrary to what might have been feared, they have not led to an atomization of knowledge into an unmanageable multitude of minutiae. Just the opposite has happened. With each gain in resolution, the diversity of life has been revealed more clearly as the multifarious expression of a single blueprint. The first unifying step came with the invention of the microscope, which led to the realization that all living beings are composed of one or more units, or cells. As more became known about the fine structure of these cells, especially after the advent of the electron microscope, their similarities became increasingly impressive. Even among cells from such diverse origins as, for example, brewer's yeast, a spinach leaf, or a human brain, common features far outnumber differences. Finally, with the recent revolutionary advances of biochemistry and molecular biology, the similarities among cells turn out to be both so close and so numerous as to render inescapable the conclusion that all known living beings are descendants from a single ancestral form. Although this form existed on Earth more than 3.5 billion years ago, the power of the mechanistic-reductionist-molecular approach has been such that we already can reconstruct in some detail and with considerable assurance the main properties of this common ancestor, as well as the principal steps whereby present-day living beings emerged from it.

These developments open the possibility of an entirely novel way of teaching biology. After a general introduction describing the biosphere and its diversity, start with the common ancestor and define the main structural and functional features that make up the basic blueprint of life. These would include proteins, nucleic acids, and the principles of biological information transfer, together with elementary genetics; enzymes, metabolism, and key mechanisms of biological energy transformation; lipid bilayers, membranes, signal transduction, and biological molecular and ionic transport, including fundamentals of bioelectric phenomena; and polysaccharides and related structural polymers. This part could include a brief survey of current ideas on the origin of life. The students would need some grounding in chemistry and energetics, though less than one might suspect. When reduced to the basic blueprint, the molecular complexity of life becomes surprisingly simple and can be rendered readily accessible.

With the common ancestor and the basic blueprint clearly traced out, current concepts of evolution can be covered and the outgrowth and ramification of the two prokaryotic kingdoms—the Archaea and the Bacteria, according to Carl Woese's most recent classification—can be outlined. This is an opportunity for introducing photosynthesis and for presenting the variegated world of bacteria, especially those that are either useful or harmful to humankind. The progressive emergence of eukaryotes—Woese's Eucarya—would then be sketched out. This would include a description of such typical eukaryotic features as the intracellular cytomembrane system and the diverse phenomena that depend on

it; cytoskeletal and cytomotor elements and organelles; and the nuclear envelope, chromosome structure, and mitotic division. Primitive protists, such as *Giardia*, can serve to illustrate these developments. Then, the endosymbiotic adoption of the prokaryotic ancestors of mitochondria and of plastids would be described, leading to the first full-fledged eukaryotic cells. The stage would be set for a historical depiction of the emergence and evolutionary diversification of the numerous species of plants and animals, including humans, that now populate the world. Stress would be put on the development of multicellular organization and on the growing complexity of intercellular communication, with a special look at the structure of the brain. Diploidy, sexual reproduction, meiosis, and recombination would be introduced, together with the principles of Mendelian genetics. Present views of morphogenesis and embryological development would be summarized. Biodiversity, biospheric cycles, planetary equilibria, ecology, ethology, even sociology could then follow.

Implementation of such a program would require a considerable effort. But I see this as mandatory for the biology of the 21st century. It is essential that the emphasis and the resulting burden on the students' intellect and memory be shifted from descriptive biology and taxonomy to the deeper kind of understanding provided by modern advances.

GROWTH, DEVELOPMENT, & DIFFERENTIATION *Living Systems Grow, Develop, & Differentiate During Their Lifetimes*	High School	2-Year College	4-Year College
Patterns of Growth	✓✓	✓✓✓	✓✓✓
Growth Rates	✓	✓	✓✓
Fluctuations in Growth Rates	✓	✓	✓
Limits in Growth	✓✓	✓✓✓	✓✓✓
Patterns of Development	✓✓✓	✓✓✓	✓✓✓
Life Cycles	✓	✓	✓
Stages of Development	✓✓	✓✓	✓✓
• Plants	✓✓	✓✓	✓✓
• Animals	✓✓	✓✓	✓✓
Differentiation	✓✓	✓✓✓	✓✓✓
Genetic Basis of Development	✓✓	✓✓	✓✓
Environmental Influences on Development	✓✓✓	✓✓✓	✓✓✓
Morphogenesis	✓	✓	✓✓
• Fundamental forms	✓	✓	✓
Tissues & Organs	✓✓	✓✓	✓✓
Form & Function	✓✓✓	✓✓✓	✓✓✓
Division of Labor	✓✓	✓✓	✓✓
Morphological Adaptation	✓✓	✓✓	✓✓
✓✓✓ = ESSENTIAL ✓✓ = OPTIONAL ✓ = NONESSENTIAL			

Table 5-5: Unifying Principle 4 with related major concepts

PRINCIPLE 5. *Energy, Matter, and Organization*
Living Systems Are Complex and Highly Organized, and They Require Matter and Energy to Maintain this Organization

All living systems consist of organized matter composed of atoms. The activities of life depend on the interactions between these atoms and their subatomic parts, so organisms and life processes conform to the principles of physics and chemistry regarding conservation and transformation of energy and matter. Living organisms differ from their physical environment. They are different because they build a nonrandom selection of atoms from the Earth's crust and atmosphere into small molecules held together by chemical bonds. Genes direct the building and maintenance of this organization. From small molecules, larger, more complex molecules are built. The most important of these are the biological macromolecules—

carbohydrates, lipids, proteins, and nucleic acids. These macromolecules have carbon skeletons, and all living systems are built of these macromolecules. Each type of organism may have macromolecules arranged in a unique way or have a few different elements added to the basic molecular structures. In some cells, macromolecules are organized into organelles, and organelles into cells. Cells are self-contained and, at least partially, self-sufficient units that conduct the basic functions of life. Each cell has a membrane at its boundary that maintains cell integrity, provides structures for self-recognition, and aids in regulating the influx and outflow of materials. Each cell also contains genetic material that determines its functions.

The body of each organism is made of at least one cell, and all cells come from pre-existing cells. Larger organisms may consist of trillions of cells arranged into a hierarchy of structures—tissues, organs, and organ systems—with each structure composed of units of the preceding level within the hierarchy. In turn, organ systems make up an individual, individuals make up populations, and populations are part of ecosystems. At each level within this hierarchy of biological organization, emergent properties appear that differentiate higher from lower levels. For instance, whereas cells exhibit characteristics of life, the macromolecules that make up the cell do not use energy, grow, or reproduce on their own. The many different types of cells reflect the myriad of life forms on Earth, but despite all the differences, organisms are unified by the processes within cells and the macromolecules of which cells are made.

The complexity and organization of a living system are maintained by its metabolism, that is, the constant intake, conversion, and utilization of energy and matter. Matter provides the chemical building blocks for the structure of an organism, and energy is needed for these materials to be assembled and this structure maintained. Without a constant supply of energy, the organization breaks down and the organism dies. In addition, the work done by organisms is powered by the energy taken in by individuals. For almost all organisms, the energy supporting the activities of life comes from the sun—either directly or indirectly. Autotrophs obtain energy from the physical environment; the most important autotrophs are plants, which use the sun's energy through photosynthesis to assimilate carbon into energy-rich macromolecules they use to grow and reproduce.

Heterotrophs rely on producers or other consumers for their resource needs, digesting food that supplies both energy and building materials. In more advanced animals, this process involves the interaction and coordination of the digestive, gas exchange, and circulatory systems. Living organisms, whether they are consumers or producers, undergo cellular respiration, during which the energy of macromolecules produced or taken in is transferred to readily usable high-energy molecules called ATP. This transfer may occur in the presence of oxygen or, in limited situations, without oxygen during anaerobic respiration. The complex chemical reactions of photosynthesis and respiration, as well as most other reactions, are controlled by specialized proteins called enzymes. Enzymes, which are organic catalysts, help to process the energy and matter taken in and help to maintain the organization of the cell.

Contemporary issues in energy, matter, and organization include the following questions. For instance, is global warming happening, and, if so, do humans contribute to the warming by using sources of energy such as fossil fuels? Which parameters are most important to know about to understand the function of the biosphere and to predict global changes? How does the amount and rate of energy and matter used by humans determine how the biosphere functions? Can models with predictive powers be developed to help us understand the effect of humans on living systems on Earth?

What are the characteristics of individuals that may allow them to obtain more energy and nutrients from the environment compared to others of their species? Some plants have a greater ability to accumulate nutrients from a given type of soil than unrelated plants, but what is the mechanism for this ability?

What controls the size and species makeup of a community? Are these characteristics related to the physical environment, the rate of flow of energy and matter through the system, the organisms that make up the community, or the interactions among these organisms?

What techniques will help reveal the organization of the cell? What are the chemical components of molecules of the cell, and how is this chemical composition related to the three-dimensional structure of the molecule?

Can the structure of a molecule be predicted based on a knowledge of its chemical makeup? Why is an understanding of the chemical organization of molecules important? In the future, will entirely new macromolecules (such as medicines) be designed and then constructed based on information gained about structure and function of known molecules?

In the essay that follows, Franklin Harold discusses the relationship between molecules such as enzymes and energy, and among all biological energy relations in general as he introduces the study of bioenergetics.

ESSAY

Bioenergetics

by Franklin M. Harold

In the world of physics, events take a predictable course. Water flows downhill, warm bodies cool, electric batteries discharge, paint peels off the wall. Things fall apart. The formal expression of these familiar truths is known as the second law of thermodynamics, which states that the spontaneous direction of natural processes tends towards the degradation of energy, the loss of potential, and the dissipation of order.

Living organisms appear, at first glance, to contradict the universal tendency of things to run down. They climb and fly, maintain body warmth, and generate electric potentials. Most remarkably, they produce molecular and anatomical order. A single fertilized egg develops into a person, composed of billions of cells functioning in concert; and on the geologic time scale, the history of life displays a record of increasing complexity and diversity. In fact, however, the productive capacities of living organisms do not conflict with the second law because they perform their labors at the expense of energy drawn from the environment. From the viewpoint of energetics, organisms are not different from a machine such as an automobile, which harnesses energy released from the combustion of fuel to run uphill. Neither organism nor machine violates the second law because when one takes into account all the energy transactions carried out by the system as a whole (e.g., an automobile plus its fuel), the overall balance is invariably negative; and that is all the law requires.

Matter and energy are probably the two most basic concepts in natural science, but whereas matter is accessible to the intuition, energy is not. In science, energy is defined as the capacity to do work. The definition of work is harder to come by and more abstract: work is most usefully understood as any process that runs counter to the spontaneous direction of events. It is work to pump water uphill and to paint the wall; likewise, it is work to accumulate solutes in the cytoplasm or to synthesize a protein molecule. Work must be done just to maintain the intricate organization of every cell and organism in the face of unremitting decay. To do work, an energy-producing (exergonic) process must be linked to an energy-consuming (endergonic) one; it is simply the tendency of the former to run down the thermodynamic hill that drives the latter uphill. Thus is electric current produced by harnessing water power with the aid of a turbine. Living things do likewise, coupling exergonic processes to endergonic work functions. To put this in less mechanical terms, energy is to biology as money is to economics: energy is the means by which living organisms purchase the goods and services they require.

Broadly speaking, bioenergetics is the study of all biological energy relations. Because the flow of energy lies at the heart of the living system, bioenergetics impinges on every level of organization within biology. For example, ecologists find that the distribution and abundance of organisms are ultimately governed by energy flow; that is why a stable ecosystem always contains far fewer wolves than deer. Physiologists deal with the energy budget when they study the rate of breathing and the mechanics of movement; energetics is the reason geese fly in a V-formation. Developmental biologists ultimately seek to understand how energy is converted into the organization of living systems. Biochemists and cell biologists explore the energy sources available to living organisms and the molecular mechanisms by which energy is harnessed to the performance of useful work.

The universe is ablaze with energy: lightning, gravitation, radioactive decay, nuclear reactions. Of this abundance of power, life (aside from minor exceptions) utilizes but two forms. Animals, and most microorganisms, rely on the degradation of preformed organic compounds. Some of these energy-producing reactions occur in the absence of air (fermentation), but the more familiar ones depend on the consumption of oxygen. Respiration by both microbes and humans may be regarded as a controlled kind of combustion. Plants, algae, and certain bacteria utilize sunlight as their chief energy source. Indeed, because the bulk of this planet's organic compounds as well as its oxygen are produced by photosynthesis, virtually the entire biosphere ultimately runs on radiant energy from the sun. The only exceptions are bacteria that live by exotic inorganic reactions such as the oxidation of ferrous iron or the reduction of carbon dioxide to methane. Regardless of their particular livelihood, all organisms must capture a portion of the energy available to them, transduce it into a usable form, and apply it to the performance of such work as biosynthesis, transport, and movement. Both the extraction of energy and its utilization are of necessity chemical processes and must take place at ambient temperature and pressure. The mechanisms by which all this is accomplished have intrigued biologists ever since the discovery of oxygen two hundred years ago.

The central issue in bioenergetics, at least in its cellular and molecular dimensions, is embodied in the phrase "energy coupling." How are the major energy-yielding pathways—respiration and light absorption—linked to energy-consuming work functions? The first conceptual unification of the field was achieved fifty years ago, when the late Fritz Lipmann recognized that adenosine triphosphate, ATP, serves as a biological energy currency much as the dollar serves as an economic currency. In a nutshell, the object of cellular metabolism is to support the synthesis of ATP, and the hydrolysis of ATP in turn pays for the performance of work. ATP itself is a reactant in processes that are energized by its consumption, such as protein synthesis. This clear and simple framework made bioenergetics comprehensible, and for many purposes it remains adequate today. The framework also focused attention on the next generation of problems, the molecular mechanisms by which ATP is generated and utilized.

A great resurgence of bioenergetics began about 1960, with the discovery that cells employ a second and very different energy currency: in many instances, metabolism is linked to useful work electrically, via a current of ions. The principle of energy coupling by ion currents found general expression in Peter Mitchell's chemiosmotic hypothesis, developed between 1961 and 1966. Following two decades of intense research and controversy, it is now quite generally accepted that the respiratory chains of mitochondria and bacteria, and also the light-harvesting apparatus of chloroplasts, are devices to pump protons (hydrogen ions) from one side of a closed membrane surface to the other; and that the return of the protons to the side from which they came (completing the circuit) drives the production of ATP by a specialized enzyme that couples proton translocation to ATP synthesis. We also have discovered that many other kinds of physiological work—most notably the transport of ions and metabolites—can be coupled to a current of protons or of sodium ions. The chemiosmotic hypothesis introduced a raft of novel and radical ideas—dual and interconvertible energy currencies, enzymic reactions that have a direction in space, and the importance of topology—that have begun to connect biochemistry to cellular architecture.

What of the future? Like its predecessor, the chemiosmotic conception of bioenergetics defined a new set of problems for a new generation of investigators. Chief among these is, how do all these molecular devices work? How are proteins designed and articulated so as to couple the hydrolysis of ATP, or the transfer of electrons, to the translocation of an ion unidirectionally across a membrane? How does light absorption by chlorophyll result in the separation of electric charge across the chloroplast membrane? And how do cells control the production of just those energy-transducing molecules required under a particular environmental regimen? Such issues now dominate the research literature, and the answers may have applications in medicine, agriculture, and industry.

I suspect that the intellectual high frontier runs, not through molecular mechanics, but through the emergent properties of complex systems. The chemiosmotic hypothesis represents a milestone along this road, for it drew attention to the directionality of biochemical processes and to the novel capacities, such as respiratory and photosynthetic ATP generation, that arise when several vectorial enzymes are housed within a single closed vesicle. Recent developments in physics open onto a still broader vista. When energy flows through a homogenous, undifferentiated system (such as a pot of water on a hot stove), the substratum may take on spatial organization; we speak of them as dissipative structures. Could it be that the foundations of morphogenesis, perhaps the very origin of life, are to be sought in this direction? I do not know, but I cherish the hope that even our utilitarian times may bring forth a few bold spirits, willing to set out along stony tracks in search of things hidden behind the ranges.

ENERGY, MATTER, & ORGANIZATION *Living Systems Are Complex & Highly Organized, & They Require Energy & Matter to Maintain this Organization*			
	High School	**2-Year College**	**4-Year College**
Molecular Structure	✓✓	✓✓✓	✓✓✓
The Chemical & Physical Basis of Biology	✓✓	✓✓	✓✓
• Scales of size & proportion	✓	✓	✓
• Oxidation-reduction reactions	✓	✓✓	✓✓
Atomic Structure/Chemical Bonds	✓✓	✓✓	✓✓
Hierarchy of Organization	✓✓✓	✓✓✓	✓✓✓
Emergent Properties within Hierarchy	✓✓	✓✓	✓✓
Carbon-based Macromolecules	✓✓	✓✓	✓✓
Cells/Cell Theory	✓✓✓	✓✓✓	✓✓✓
• Organelles	✓✓	✓✓✓	✓✓✓
Matter	✓✓	✓✓✓	✓✓✓
Assimilation/Ingestion/Digestion	✓✓	✓✓	✓✓
Transport of Materials	✓✓	✓✓	✓✓
• Diffusion/active transport	✓✓	✓✓	✓✓
Membranes	✓✓	✓✓	✓✓✓
Digestive, Gas Exchange, & Circulatory Systems	✓✓	✓✓	✓✓
• Blood	✓✓	✓✓	✓✓
Nutrients	✓✓	✓✓	✓✓
Energy & Metabolism	✓✓✓	✓✓✓	✓✓✓
Enzymes	✓✓✓	✓✓✓	✓✓✓
ATP & Energy Transformation	✓✓✓	✓✓✓	✓✓✓
Autotrophs	✓✓	✓✓	✓✓
• Photosynthesis	✓✓✓	✓✓✓	✓✓✓
• Chemosynthesis	✓	✓	✓
Heterotrophs	✓✓	✓✓	✓✓✓
• Aerobic respiration	✓✓✓	✓✓✓	✓✓✓
• Anaerobic respiration	✓✓	✓✓	✓✓✓
✓✓✓ = ESSENTIAL ✓✓ = OPTIONAL ✓ = NONESSENTIAL			

Table 5-6: Unifying Principle 5 with related major concepts

PRINCIPLE 6. *Maintenance of a Dynamic Equilibrium*
Living Systems Maintain a Relatively Stable Internal Environment Through Their Regulatory Mechanisms and Behavior

An organism requires a relatively stable internal environment to obtain and use resources efficiently, to grow, and to reproduce. Organisms, however, are confronted by a constantly changing environment—one that varies on a daily, seasonal, and yearly basis. Many of these changes can be lethal, and living organisms therefore must have the capacity to adjust to or avoid these fluctuations. They must regulate the amount and type of materials they take in and the ways in which they use these materials. They also must avoid extreme fluctuations in internal temperature, maintain water balance, and eliminate waste products through excretion. In his essay on homeostasis and regulation, Knut Schmidt-Nielsen discusses mechanisms living organisms use to maintain constant conditions and to resist the effects of environmental changes—important aspects of maintaining a dynamic equilibrium.

Homeostasis is the maintenance of a relatively stable internal environment within a varying external environment. To maintain that stability, an organism first must sense a change and then regulate its physiological activities to adjust. The ability to regulate is made possible by feedback mechanisms that sense changes inside and outside the organism and then respond to them. In response to environmental stimuli, organisms may, with different degrees of ability, regulate their body temperature and the chemical composition of their cells, and defend themselves against foreign material and organisms. For example, the symptoms of disease are indicators that a host is regulating its internal environment after invasion by a disrupting pathogen. These highly specific defensive reactions are part of an animal's immune response. Some plants also produce chemicals that confer disease resistance to themselves because the chemicals are toxic to certain pathogens.

Organisms also may respond to the environment by switching chemical reactions on or off, transporting materials from one place to another, moving, or experiencing differential growth and reproduction under the influence of hormones, enzymes, or other chemicals. Hormones are organic molecules that regulate the function of specific tissues or organs. While growing and developing, a living system may adjust to changes in its environment—both inside and outside itself—through a combination of behavioral, physiological, and morphological changes.

All organisms respond to stimuli, or environmental cues. Plants, for example, may demonstrate tropisms in response to light and gravity. Animals respond to their environment through their behavior. Behavior is an individual's response to an environmental stimulus, a set of actions determined in part by the individual's genes, and the physical manifestation of an internally orchestrated process. A behavioral response usually involves some action that must be coordinated among the activities of many cells. Such coordination depends on communication between the sensory receptor cells that receive stimuli from the environment and the motor response cells such as muscles that carry out the behavior. Communication may require hormones produced by the endocrine system and, in organisms with well-developed nervous systems, involves neurons—highly modified cells that transmit information from sense receptors to a central nervous system and then to cells that respond. Such complex organisms also possess sophisticated means to sense the environment, bodily structures that allow quick reactions, and a brain to coordinate incoming stimuli and outgoing responses.

An organism's behavior encompasses a wide range of activities in response to particular stimuli important for its survival and reproduction; there is an adaptive value to the behavior. How an organism moves, when it does so, where it goes, and what it does when it gets there all are aspects of an organism's behavior. Such behavior may be innate, resulting in a complex, predictable response to a first-time encounter with a particular stimulus. As an animal matures, behavior related to reproduction may develop. More complex animals also may learn and, therefore, modify their behavior based on some of the experiences they have accumulated. Because an animal does not live an isolated existence, it may exhibit territoriality against its rivals and altruism toward its gene-sharing relatives. It also may be part of a dominance hierarchy within the group to which it belongs, another behavior of individuals within certain animal societies.

A few **contemporary issues in the maintenance of a dynamic equilibrium** include the following questions. For instance, how do plants measure changes in the environment? Some plants use cold temperatures as a cue for the initiation of developmental processes that occur in the spring or early summer, but how do they measure temperature, how do they determine the length of a cold period, and how do they translate this information into a developmental action?

What discoveries will the study of AIDS lead to that are unrelated to the cure or control of AIDS itself? What can be learned about the human immune system and its normal ability to inhibit and destroy pathogens? What role do the different cells of the body's immune system play in the duplication, control, and concealment of the AIDS virus?

How do cells communicate with other cells within the body of an organism? How will an understanding of cell communication help in the treatment or cure of certain human cancers and forms of mental illness that result from abnormal cellular regulation?

How does the brain control the movement of the body? How are the different systems within the brain organized, and how do they function to direct sensory perception, language and motor functions, emotions, and thinking abilities in humans?

How do people learn, and can learning be improved by understanding the action of the nerve cells? In what form are memories and information stored? What role do proteins, genes, and neurons play in memory? Is learning in nonhuman animals similar to learning in humans?

How do nerve cells, the processing and signaling units of an animal's brain, direct behavior? How are the physiological events in the nerve cells related to the transmission and processing of information? What are the differences between the many types of nerve cells distinguished within individual animals?

Is there a molecular basis to neurological disorders, and, if so, can this information be used to diagnose the problem? Neurological diseases that have a known genetic origin include Huntington disease, a form of Alzheimer's disease, and neurofibromatosis, but how do genes, the brain, and behavior interact in these disorders?

James Gould outlines the discipline of animal behavior and the biological basis of behavior in animals in the essay that follows *Regulation and Homeostasis*.

ESSAY

Regulation and Homeostasis

by Knut Schmidt-Nielsen

Living organisms tend to maintain a relatively constant state that is optimal for survival. This constancy is well regulated and is called *homeostasis*. Deviations from the usual state or composition of the organism are met with corrective action. Deviations that are too great cause serious difficulties and eventually death. Homeostasis is important at all levels, from the single cell through the level of organs to the entire organism. In ecology the term applies to the self-regulating maintenance of population densities.

Living organisms have elaborate mechanisms to resist changes and maintain constant conditions. Regulatory mechanisms respond to deviations or departures from the normal; corrective action returns the system to the normal range. Mechanisms to maintain constancy involve feedback control. A familiar example of feedback control is the heating system of a house that is regulated by a thermostat. If the temperature falls below the desired temperature (the setpoint), it serves as a signal that causes corrective action (more heat),

thus restoring the system to the desired temperature. When the deviation (in our example a decreased temperature) is offset by a corrective action in the opposite direction (increasing the temperature), it is called negative feedback.

A biological example of feedback control is the maintenance of a relatively constant body temperature. If our body temperature increases, we normally increase heat loss by sweating; if the body temperature drops, we respond by shivering to increase heat production. The responses thus tend to maintain the body temperature at the normal level.

An example of homeostasis in physiology is the aquatic animals that maintain salt concentrations in their bodies different from those in the surrounding water. Marine invertebrates maintain the same total osmotic concentration as in the surrounding sea water, but the concentrations of various salts in their blood invariably differ from those in the surrounding water. Similarly, the concentrations of various salts inside the cells differ from those in the blood and extracellular body fluids. These differences are characteristic of each animal species, and their lives depend on the maintenance of these differences. Marine fish, in addition to maintaining internal concentrations different from those in sea water, maintain their overall osmotic concentration at about one third of that in sea water.

Fresh water has very low concentrations of salts, and both invertebrates and vertebrates that live in fresh water (fish, some amphibians and reptiles) maintain internal concentrations that far exceed those in the very dilute water. The maintenance of differences between the organism and the surrounding water depends on elaborate control mechanisms, a subject that generally is referred to as *osmoregulation*.

Many of the processes that serve to maintain constant conditions in higher vertebrates, and especially in mammals, are reasonably well known and understood. Such functions include the maintenance of constant body temperature, constant heart and breathing rates, constant blood pressure, and constant level of blood sugar. For example, the mechanisms involved in the maintenance of a constant water content of the body, in spite of changes in water intake, are well known. If water intake is restricted, the kidneys respond by reducing the volume of urine, eliminating waste products in high concentrations with a minimal use of water. If water intake is large, the kidneys respond by producing large volumes of dilute urine. These changes in renal function are under hormonal control; if there is water shortage, the pituitary gland produces antidiuretic hormone, ADH (also known as vasopressin); if there is a surplus of water to be eliminated, the production of the hormone ceases and urine volume increases. In mammals it is the retention and conservation of water that is promoted by the antidiuretic hormone ADH. In other animals it may be water elimination that is augmented by a hormone. Consider a mosquito, which in a single blood meal may ingest twice its body weight, a load

that greatly impairs its flying and maneuvering ability unless it is rapidly eliminated. A blood-sucking mosquito, before it has even completed its meal, begins to urinate at a high rate. The sudden increase in urine production is regulated by a diuretic hormone that stimulates secretion of fluid by the Malpighian tubules.

We are familiar with the responses to acute changes in physiological demands such as while exercising. The increased oxygen demand during exercise is met by increased rate and depth of breathing, increased blood flow to the working muscles, and increased heart rate (plus a moderate increase in stroke volume). These changes are carefully regulated to meet the demands and, when exercise ceases, the rates return to the resting levels.

Also familiar to us are the slow changes that occur in response to increased physical demands on the organism during athletic training. The changes involve well controlled reorganization and remodeling of physiological systems in response to the increased demands. Muscles subjected to increased use respond with growth and with less obvious changes, such as increases in number of mitochondria, in Krebs cycle enzymes, in cytochromes—responses that augment the performance of the muscles. Training also augments the heart's capacity for pumping blood and the lungs' capacity for oxygen uptake.

If the bones are subjected to long-term changes in forces exerted on them, they undergo well understood changes. Bones are far from dead and inactive mineral tissue; living cells in the bones (osteoclasts and osteoblasts) respond to changes in the forces exerted on the bones by resorption and deposition of mineral substances (crystals of inorganic calcium phosphates). A major question that today we are unable to answer fully is how the forces on the bones are communicated to the cells responsible for the remodeling of the bones.

The immediately preceding sentence points to an unsolved problem that is of the greatest importance. While the mechanisms involved in the maintenance of steady state responses to short-term changes such as exercise are reasonably well understood, the mechanisms involved in the regulation of long-term changes are not. For example, consider the ability of the liver to regenerate. A mammal can tolerate surgical removal of a large portion of its liver. The remaining liver will begin to regenerate the lost tissue. The growth of new tissue continues until the liver reaches its initial size, and then it ceases. Little is known about the signals that initiate the regeneration, except that certain growth-promoting substances are involved, and nothing is known about the signals that make the growth cease when the regenerating liver reaches the "correct" size.

Many related problems pertain to other areas of the growth and development of an organism. We know, for example, that the normal growth of a human infant depends on the release of pituitary growth hormone. The sudden spurt in growth of the human male in early puberty is initiated by the influence of gonadal hormones and their effect on the production of growth hormone, but what in turn regulates this increase in production of male hormone and makes it cease when the body has reached the "correct" size is poorly understood.

Consider a fertilized mammalian egg that has undergone two divisions and hence has reached the four-cell stage. One of the four cells can be removed and then transplanted into the uterus of a suitably prepared female of the same species. This cell, only one fourth of the size of the original egg, is now capable of growing into a normal embryo that eventually develops into a normal, full-sized fetus that is delivered at term. We do not understand how a single cell that represents only a fraction of the mass of the fertilized egg grows into a normal fetus, what signals control its growth to the right size, and what makes growth cease at the exact correct size. How does the organism "know" what its right size is?

Today, we are seeking answers to these and similar questions pertaining to the long-term control of growth and development, as well as long-term remodeling in response to changing demands on the organism. Our current knowledge is inadequate, and the many unsolved problems in these areas are foremost in the minds of many contemporary biologists.

ESSAY

The Biological Basis of Behavior

by James L. Gould

The study of behavioral biology serves two key roles in a biology curriculum: it exposes students to some of the most interesting and relevant phenomena in modern biology, and it provides an excellent bridge between material on animal physiology (particularly the nervous system and sensory physiology) and the study of evolution and ecology. This bridge, properly handled, helps unify two major areas of the biological sciences.

Behavioral biology is the study of behavior at three levels: mechanisms, organization, and evolution. The discipline can be subdivided into neuroethology, ethology, and behavioral ecology, each of which draws from all three levels, but to very different degrees. After briefly summarizing these divisions, I will describe in more detail the essential contribution each makes to a broad understanding of behavior.

Neuroethology concentrates on behavioral mechanisms. These include the sensory apparatus with which animals monitor themselves and the world around them, the neural processing that edits and enhances those data, and the circuitry that orchestrates responses. Though neuroethology is primarily concerned with mechanisms, it draws upon and sheds light on the other two levels: the circuits it has elucidated are the basis of behavioral organization, and cross-species comparisons of sensory and processing adaptations are among the most powerful tools in understanding behavioral evolution.

Ethology has traditionally concentrated on the behavior of individual animals as they respond to internal changes and external cues from members of their own or other species, or from the environment. Often referred to as the study of instinct, ethology deals with how behaviors such as hunting/foraging, navigating, communicating, and learning are organized. Ethologists are particularly interested in how stimuli are used to direct and trigger behavior, how the behavior itself is "programmed" to take into account a range of contingencies, how complex, multi-step behavior (e.g., nest building) is organized, and how competing needs (such as hunger and courtship) are balanced and logical decisions between behavioral alternatives are made.

Behavioral ecology concentrates on how each behavioral sequence has evolved, as opposed to the many alternatives that are possible—particularly those observed in related species. As a result, behavioral ecologists look closely at a species' ecological niche for clues to the adaptive logic of behavior. The approach draws heavily on cross-species comparisons and a weighing of the selective costs and benefits of behavioral alternatives. Much of behavioral ecology is concerned with social interactions and the evolution of social systems.

Behavioral biologists approach their studies from the bottom up (neuroethology → ethology → behavioral ecology) or from the top down, though the former is more satisfactory. Neuroethology can build on the knowledge of how neurons work and sensory organs operate. This link between the nervous system and actual behavior is an essential part of the bridge behavioral biology provides, strengthened further by the use of cross-species comparisons that draw on the effects of natural selection in shaping sensory processing. For instance, the processing underlying sound localization is an example rich in illuminating diversity: there are three basic types of ears (particle-motion, pressure-gradient, and pressure-difference), each with its own set of advantages and drawbacks; each species must use its ears to identify important auditory stimuli, localize the source, and trigger an appropriate response. The potentials and limitations of each kind of ear have clear effects on the processing strategy that has evolved and the behavior it serves.

Another important concept of neuroethology is that a few, relatively simple types of neural circuits accomplish most kinds of information processing and orchestrate most sorts of elementary innate behavior. One good set of examples can be drawn from work on the sea hare, *Aplysia*, where the circuit involved in gill withdrawal illustrates the wiring of a stereotyped response, while those circuits that generate feeding and blood flow demonstrate how repetitive behavior is produced. In sum, an exposure to the basic concepts of neuroethology provides a tangible, mechanistic reality to the otherwise mysterious phenomena of animal and human behavior, much as the study of replication, transcription, translation, and meiosis establishes a firm mechanistic basis for the phenomenology of Mendelian genetics.

Ethology logically begins by looking at the organization of simple innate behaviors. One of the most illuminating examples is the egg-recovery and imprinting behavior of geese and certain other ground-nesting birds. A goose must be able to recognize eggs when it is incubating and know how to retrieve them when one has rolled out of the nest. As behavior like this is described analytically, each of the major themes of ethology unfolds: *sign stimuli* (only certain features of the egg are important, so many things that are not eggs—soda cans, for instance—will be retrieved); *motor programs* (the retrieval behavior is innate, so it will continue to completion even if the egg is removed while the goose is extending its neck reaching for it; at the same time, the behavior has a built-in capacity to modify itself so that the bill, with which the egg is rolled back, is automatically moved from side to side to compensate for unevenness in the surface over which the egg is being moved); *drive* (egg-retrieval behavior appears shortly after nesting begins, but before any eggs are laid, and persists until about a week after the chicks have hatched); and innately guided *learning* (geese are unable to learn anything about their own eggs, but the chicks imprint on their parents, recognizing them as suitable objects to follow, commit to memory, and flee to when in danger, on the basis of specific sign stimuli).

Most of the work of ethology has involved attempting to understand how these basic mechanisms are used to accomplish other behavioral tasks. Perhaps the most important and unifying set of behaviors is the set that involves learning: the basic circuitry of associative learning is known, and the phenomenon, beyond its intrinsic interest, illustrates the seamless integration of instinct and learning, two concepts often considered mutually exclusive.

The understanding of how individual behavior can be generated leads smoothly into the questions of behavioral ecology: how did this particular way of responding to the challenges posed by predators, prey/food, the habitat, and other members of the species evolve, rather than some other strategy? Behavioral ecology introduces and draws on basic ecological and evolutionary principles like niche (an animal's "occupation"), competition, and natural selection as a net sum of costs and benefits. Behavioral ecology is particularly powerful at introducing the concept that selection operates on individuals rather than, as is commonly thought, on groups or species. The most illuminating examples come from cross-species comparisons of social systems, examining the costs and benefits of sociality (competition, interference, and higher rates of disease transmission, for example, versus cooperative defense or hunting) and the behavioral mechanisms that make sociality work (such as dominance hierarchies and intraspecific communication). In many cases it is possible to demonstrate the existence (and adaptive value) of alternative strategies in a species; because there is in many cases no perfect response for all occasions, an animal is thus able to choose the strategy that is most likely to benefit an individual in its position.

Throughout any study of behavioral biology, scientists find it useful to consider the implications for human behavior. Our sensory blindnesses and the biases in perspective that result from our neural processing become evident as one learns about neuroethology and orientation. The operation of ethological mechanisms (sign stimuli, motor programs, drive, and selective learning) are clearly evident in many behaviors; language acquisition is perhaps the most dramatic. A consideration of the behavioral ecology of our species' ancestors provides much food for thought in interpreting our contemporary social arrangements and behavior.

Although the questions facing researchers working at every level in modern behavioral biology can be made clear in any discussion of animal behavior, many have their greatest intellectual impact when placed in the context of human behavior, and incidentally provide links to psychology, sociology, and anthropology. The most pressing questions include the relationship between the "feature detectors" of neuroethology and the sign stimuli of ethology, the degree of complexity possible in innate recognition, the degree and nature of sensory calibration during development and early life, the workings of drive, the nature of memory storage and retrieval, the organizational basis of motor (operant) learning, the organizational basis of associative (classical) learning, the basis of human perceptual and learning disabilities, the behavioral rules that underlie choice behavior, how the action of selection on neuroethological mechanisms actually produces behavioral change, the interaction between genetic heritage and cultural learning, the nature of insight and creativity, and the relevance of behavioral understanding to managing and conserving species.

MAINTENANCE OF A DYNAMIC EQUILIBRIUM Living Systems Maintain a Relatively Stable Internal Environment	High School	2-Year College	4-Year College
Detection of Environmental Stimuli	✓✓	✓✓	✓✓
Receptors	✓	✓	✓
Movement	✓✓	✓✓	✓✓
Effectors/Muscular Systems	✓✓	✓✓	✓✓
Skeletal/Support Systems	✓✓	✓✓	✓✓
Homeostasis	✓✓✓	✓✓✓	✓✓✓
Feedback Mechanisms	✓✓✓	✓✓✓	✓✓✓
Nervous Systems	✓✓	✓✓	✓✓
• Brain	✓✓	✓✓	✓✓
Endocrine Systems	✓✓	✓✓	✓✓
Hormones	✓✓	✓✓✓	✓✓✓
Temperature Regulation	✓✓	✓✓	✓✓
Water Balance/Excretory System	✓✓	✓✓	✓✓
Health & Disease	✓✓✓	✓✓✓	✓✓✓
Fitness	✓✓✓	✓✓✓	✓✓
• Human nutrition & problems	✓✓✓	✓✓✓	✓✓✓
Immune System & Response	✓✓	✓✓✓	✓✓✓
• Pathogens & response	✓✓	✓✓✓	✓✓✓
Human Diseases	✓✓✓	✓✓✓	✓✓✓
• Medical & public health issues	✓✓✓	✓✓✓	✓✓✓
• Biomedical technology	✓✓	✓✓	✓✓
Human Genetic Disorders	✓✓	✓✓	✓✓
Behavior	✓✓	✓✓✓	✓✓✓
Tropisms	✓	✓	✓
Communication	✓	✓	✓✓
Animal Behavior	✓✓	✓✓	✓✓
• Instinct	✓✓	✓✓	✓✓
• Learning	✓✓	✓	✓
• Human Behavior	✓✓✓	✓✓✓	✓✓
− Determinants	✓✓	✓✓	✓✓
− Drugs & their effects	✓✓✓	✓✓✓	✓✓

Table 5-7: Unifying Principle 6 with related major concepts

Conclusion

One goal of biological literacy is the acquisition of useful knowledge that serves as a foundation for learning additional information and that can be applied to an individual's decision-making process. The acquisition of biological content should not be the only goal of any biology program, but there are major concepts that are essential for a biologically literate individual to understand and appreciate. From the discussion of the unifying principles of biology and their associated concepts presented in this chapter, we have identified the following twenty concepts as an essential part of a biological education for high school, community college, and four-year college students. An instructor should view this list as a guide to exclude the teaching of optional and nonessential concepts and to expunge the idea that one must cover an entire textbook, or even very much of it, during a course. The twenty major biological concepts and their related unifying principles are:

Evolution
1. The patterns and products of evolution, including genetic variation and natural selection
2. Extinction
3. Conservation
4. Characteristics shared by all living systems
5. Overview of biodiversity, including specialization and adaptation demonstrated by living systems

Interaction and Interdependence
6. Environmental factors and their effects on living systems
7. Carrying capacity, and limiting factors
8. Community structure, including food webs and their constituents
9. Interactions among living systems
10. Ecosystems, nutrient cycles, and energy flow
11. The biosphere and how humans affect it

Genetic Continuity and Reproduction
12. Genes and DNA, and the effect of interactions between genes and the environment on growth and development
13. Patterns of inheritance demonstrated in living systems
14. Patterns of sexual reproduction in living systems

Growth, Development, and Differentiation
15. Patterns of development
16. Form and function

Energy, Matter, and Organization
17. Hierarchy of organization in living systems
18. Metabolism, including enzymes and energy transformation

Maintenance of a Dynamic Equilibrium
19. Homeostasis, the importance of feedback mechanisms, and certain behaviors
20. Human health and disease

Biology is a scientific discipline; biologists are scientists investigating the part of the natural world that contains living systems. Biology can be divided into two parts: (a) biological concepts that have been developed through (b) the process of scientific inquiry. As discussed in this chapter, biological concepts are organized around three key elements: (a) the unifying principles of biology; (b) levels of organization; and (c) the diversity of living systems. These elements are represented by the interacting matrix shown in Figure 5-1. All biological knowledge is found in the intersections of these three elemental sides of the figure. For instance, the intersection of (a) *energy, matter, and organization,* of (b) *plants,* at the (c) *cellular* level encompasses many questions related to photosynthesis. Or, a scientist may investigate questions related to *patterns and products of change* in *animals* at the *population* level as he or she studies human population genetics. These three interacting sides reflect one way to organize the vast amount of biological information.

Biological information has been obtained through a systematic study of the natural world. This process of investigation is represented by the three interacting sides of the matrix shown in Figure 5-2. One side indicates a scientific method, a formalized model of the processes of science. Although scientists usually do not follow the components of a scientific method in a rigid manner, each of the steps contributes to a logical process of investigation.

The second side of Figure 5-2 represents different ways that scientists may approach their research problem. In the early stages of investigation, one describes a natural phenomenon and then conducts experiments based on the questions generated during observation. Some scientists may investigate phylogenetic relationships of particular organisms, study the components, regulation, and change involved within a system, or try to develop a model of a natural system. Still other investigators may try to determine the cause of some observed phenomenon, or the relationship between the function of a particular structure and the morphology or anatomy of that structure.

The third side of this matrix includes factors that affect and are part of scientific investigations, including such influences as the historical development of a particular subject and the technology that might be used to answer basic biological questions. Technology is not synonymous with science, although today they are closely intertwined. Scientists use technology to discover explanations of the natural world and to communicate this knowledge to others. Technology is important in solving human problems and in adapting the environment to human needs. Together, the three sides of this matrix reflect the processes of scientific investigations.

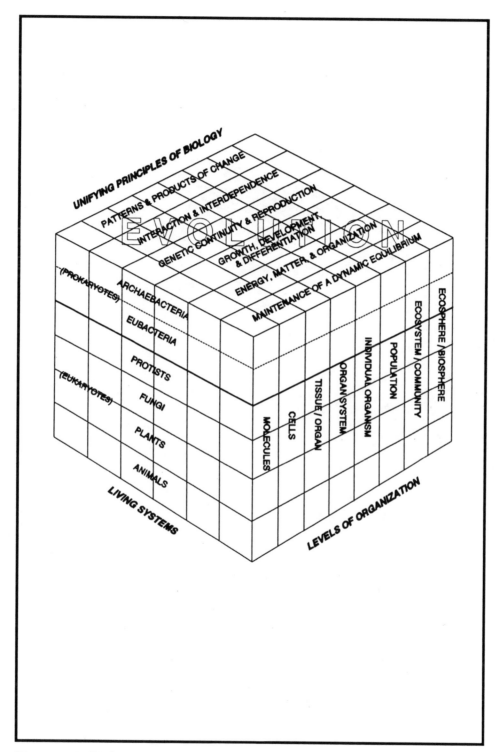

Figure 5-1: Biology as a scientific discipline: Biological concepts

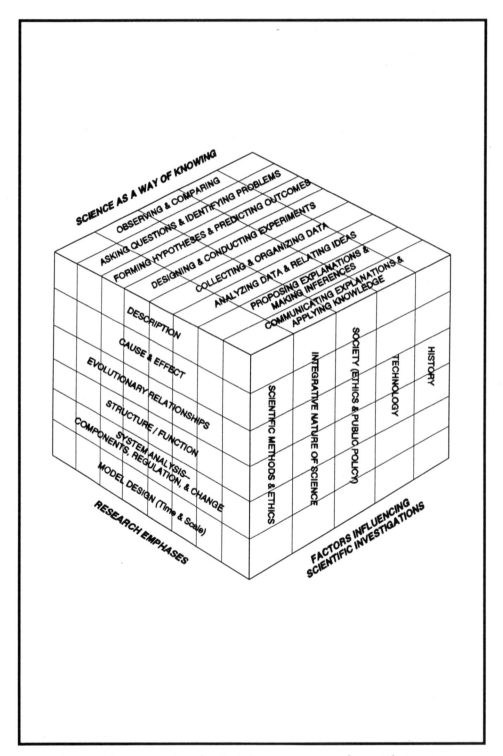

Figure 5-2: Biology as a scientific discipline: Processes of scientific inquiry

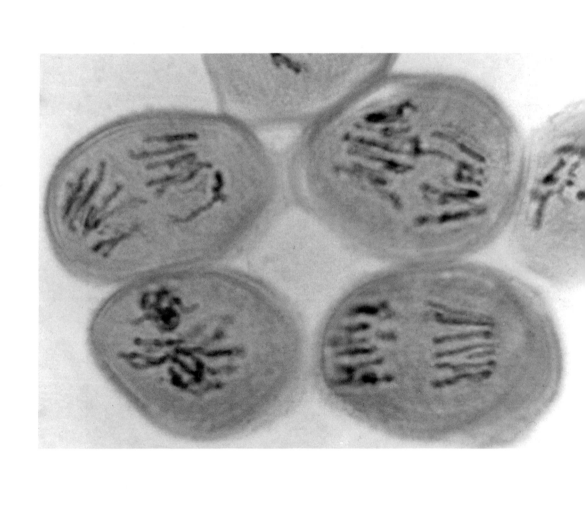

6

Designing a Contemporary Biology Program

This guide identifies several key components of a biology program designed to develop biological literacy in students. In planning a program in biology, there are many components to consider. It becomes readily apparent that each component adds another layer of concerns onto the complex outline that eventually becomes the learning environment of the classroom. The difficulty, of course, is to decide which components to use and then how to juggle them all at once.

The major consideration in program design is, of course, the student. Although high school students differ from college students in their background and development, post-secondary instructors should not be precluded from using similar themes or focusing on the same concepts to which students are exposed in high school. In fact, college biology classes should build on the conceptual and pedagogical foundations established in high school because biological literacy is a life-long process requiring students to build on the knowledge, values, and skills they gain in earlier classes. One might expect, however, that college students would demonstrate a comparatively higher level of analysis regarding concepts, themes, and activities and would therefore be able to study more complex and controversial problems than those younger students could study. At all levels of education, students should be able to investigate a biological question or a biosocial problem using the processes of scientific investigation and to understand biological concepts within the framework of the unifying principles of biology. The focus on investigative and critical thinking skills and on the organizational power of the unifying principles then addresses the complaint registered by many students who claim they have studied the same biological factual knowledge—for example, the name of the bones in the human arm—in every biology class they have taken. By selecting problems for investigation, students can discover biological information that is unfamiliar to them while using methods that are familiar.

The components chosen for a biology program should be appropriate to the development of the students. High school teachers should recognize the needs and interests of adolescents. Community college instructors must recognize the needs and interests of a variety of students, including reluctant and adult learners, while preparing students to become participating citizens. Four-year college and university faculty should recognize the needs and interests of young adults who will not be biologists or enter health-related professions but who are making daily decisions related to biology, health, and other science-related issues. Nonmajors courses often skimp on their presentation of the activities and results of science that are needed both by the student who *might* want to become a scientist and by the student who, as a citizen, needs to know more about what science is and how it operates (Project Kaleidoscope, 1991).

We suggest that you begin designing your program by asking yourself these three questions: (1) What are the goals of the program for me and my students? (2) How do I intend to reach these goals? and (3) How will I know I have achieved these goals?

What Are Your Goals?

In establishing goals for your program, you should first consider your students. What do you want them to know, value, and be able to do after completing your biology program? To help your students attain a structural or multidimensional level of biological literacy (Chapter 3), we recommend the following:

- Students should experience science as a process. This means that each program should include investigative experiences for all students. These experiences may be a set of laboratory investigations, independent research projects, or inquiry-oriented activities that emphasize the qualitative and quantitative aspects of biology.
- Focus on a limited number of major biological concepts that are linked by the unifying principles of biology.
- Include varied instructional strategies and appropriate assessment procedures that challenge students' higher-order thinking and reasoning skills.
- Provide opportunities for student discussion, explanation, and linkage of biological principles and concepts.
- Present information that is relevant to the personal lives of students and provide opportunities for students to apply this knowledge in the resolution of authentic problems.
- Assess the ability of students to use the processes of science and to solve problems at their level of understanding.

One way to integrate experiments, demonstrations, lectures, and discussions is to incorporate all such strategies into a regularly scheduled class period rather than creating the typical dichotomy of lecture and laboratory. Under this structure, each class period will include several activities, for example, shifting back and forth from hands-on activities, to discussions of concepts, to a lecture on related topics.

A goal for each instructor might be to improve his or her teaching style, and one way to assess instructional skills is to review widely acknowledged characteristics of exemplary biology teachers. For example, excellent instructors:

- establish a learning community in the classroom, that is, they cultivate an atmosphere in which discussion is encouraged—e.g., they divide a large class into small groups, or they use cooperative-learning techniques;
- actively engage students in biology through the use of activities such as laboratories and field trips;
- make connections to their students' world by using concrete activities to teach abstract concepts and by identifying local and human issues that concern students;
- make decisions in terms of their students, for example, they are flexible enough to accommodate unanticipated student concerns, interests, or discussions;
- use a variety of instructional resources and strategies;
- use analogies and examples;
- establish positive personal relations with their students by listening, understanding their problems and concerns, and caring about their progress in the class; and
- measure the success of their program against the knowledge, skills, and values demonstrated by the students.

One must reduce or eliminate activities that do not promote higher levels of biological literacy while focusing on activities that do. Because many high school and college biology courses are taught using a lecture format, we first address this widespread instructional strategy and then discuss reaching the goal of biological literacy through implementation of problem-solving activities.

Improving Lecture Courses

This report advocates that instructors move from a lecture-only format to one that uses a variety of activities. If a gradual restructuring of the biology program is desired, then one or more new instructional strategies (Chapter 4) could be added each term (while reducing the lecture component), eventually resulting in a course with varied instructional and assessment strategies. The following suggestions may help a lecture course evolve into one that actively engages students.

The greatest change in a lecture course comes from the addition of a laboratory or other investigative experience (or the improvement of existing laboratories). Ten minutes is near the upper limit of comfortable attention students give to lecture material, while the attention span in an investigative laboratory is far longer (Project Kaleidoscope, 1991). Also, because the essence of science is process, method, and practice, it is critical that students have ample opportunities to conduct scientific investigations.

Teacher-directed and student-directed group discussions during which students explain concepts and develop critical thinking skills could be added to or replace lectures. The instructor acts as the facilitator of discussion or asks directed questions based on assigned research articles, historical vignettes, or newspaper or magazine reports. Other components that could be added to a lecture course are writing assignments or projects *outside* of scheduled class time—in addition to reading. For instance, short assignments might require students to summarize their understanding of information presented in lecture. Student research projects that are independently designed and independently conducted can be completed in the laboratory (if time and space permit) or at the student's home.

One can begin each lecture or major section of a lecture with an intriguing question that underlies all of the information presented. By asking such an icebreaker question, and waiting, some kind of response from students is assured. This allows students to develop the habit of answering questions. During lectures, instructors should schedule pauses for student questions, discussion, and assimilation of ideas. Many speakers develop an interactive lecture style during which they break from speaking and ask open-ended questions. These may be from students who anonymously deposit questions about lectures or materials into a box before or after class and that are dealt with as soon as possible. Guest speakers and presenters also break up the monotony of a single lecturer, as does a debate between students on a topic of interest to them. Finally, if students have an opportunity to read lecture notes before class, many more of them actually may *listen* during the lecture and *understand* what is presented rather than just copying words. The use of interactive audiovisual materials and non-verbal visual props to illustrate lecture material (such as flowers that are passed around the room) can keep students interested. Traditional audiovisual programs can be made more interactive by first showing part of the program, breaking for questions and discussion, and then returning to view more of the program.

Small-group work is an effective alternative to lecture. The class may be divided into small groups to which a variety of activities are assigned. Such work is facilitated by scheduling the course in a room that will allow a large class to be divided into small, manageable discussion groups. Each group of students may attempt to answer a question posed by the instructor or investigate a biological problem that may require work outside of class.

Whatever strategy chosen, moving away from a lecture-only course provides students with a richer and more interesting experience in biology and increases the likelihood that they will become more biologically literate.

Moving Students Along the Continuum of Biological Literacy

Structural and multidimensional biological literacy are appropriate levels for your students to reach. It is important to remember that while someone may be multidimensionally literate about one concept of biology, he or she may be only nominally literate about other concepts. It is important to focus on a few essential concepts you want your students to understand, and to spend more time developing these concepts.

In an inquiry-based biology program, students construct their own, correct explanations of biological structures, concepts, and natural phenomena. For instance, instead of simply telling students the function of a cotyledon, allow them to conduct an investigation on seed germination in which they remove various portions of the cotyledons and follow the subsequent growth and death of the seedlings. Such activities necessarily require an instructor to cover fewer topics in a year, but students will be able to build a better understanding of the information obtained from such hands-on activities as opposed to hearing or reading about them.

Student interest in biology may be increased by using hands-on, exciting laboratories and activities, and by helping students see the relevance of the subject to their own lives (Project Kaleidoscope, 1991). For example, to engage students in a learning activity, the selection of a local land-use issue, such as where to put the new landfill, can show students that what they are learning in class (e.g., population growth, resource use, and carrying capacity) has relevance and practical application outside of the classroom. In addition, newspaper and magazine articles, television newscasts, and reports from other media can illustrate that the concepts dealt with in class are related to important national, international, and personal concerns.

One may help students achieve multidimensional literacy by several different routes. Students may continue their study of a subject if their personal interest remains high. For instance, if a student or a close friend or relative contracts or inherits a disease, that student may become engaged in the study of the biology of that disease. For those students who are curious about any subject, an instructor can point to additional readings and research that may address unanswered questions and that could lead to open-ended independent investigations.

Other students may continue study of a biological concept if an instructor presents a problem that affects them directly or if the students find themselves confronted with a personally relevant local, regional, or global conflict. If necessary, instructors can initiate this process by posing problems or identifying real conflicts within the community that stimulate interest. Students then can be challenged to apply their knowledge to the problem's resolution. By presenting an unresolved local public issue, you also give students the opportunity to become directly involved in a multidimensional investigation that is relevant, in which they practice the processes of scientific investigation, *and* in which they learn biological concepts and the connections to other concepts and disciplines. These activities help blur the distinction between pedagogy and content and the distinction between classroom and laboratory, and help demonstrate the property of connectedness in science. Connections give the learner something to think about, and lessons lacking connections are meaningless, rote, and authoritarian (Project Kaleidoscope, 1991).

An issue-centered or problem-based approach to teaching as described above organizes the curriculum around opportunities for students to practice and develop critical thinking, analysis, and decision-making skills. In this approach, students may work on a problem for several weeks or months and obtain and use knowledge and skills from several different disciplines (O'Neil, 1992). For instance, students may be presented with a problem about a fictitious pregnant woman and her developing fetus that medical tests indicate may be anencephalic. The parents of the fetus are interested in the possibility of using the fetus as a donor of organs or tissue, but have some reservations based on sanctity-of-life issues. Students then must consider the implications of the medical tests for the parents, the fetus, and society, and they must determine what they need to find out to make appropriate recommendations to the parents. Students are pushed to explain and support their statements, and are directed by the instructor to think critically and to develop ethically defensible positions during their investigation. The investigative process leads students through problem identification, hypothesis building, research, synthesis of ideas, and finally to decision making. Students may be assessed on the scope and accuracy of the information they provide and fictitious advice they give, the clarity of their presentation, and their response to questions (O'Neil, 1992). Such long-term investigations will help students develop higher levels of biological literacy.

Evaluating Your Personal Program

Developing biological literacy in students requires an analysis of current program practices, long-term goals, and consideration of how one might move toward these goals in the immediate and near future. After reviewing information in this guide, completion of a table such as the one illustrated in Figure 6-1 should help you develop your biology program. To develop a table for yourself, identify all the components of the program using information in the previous chapters, and evaluate your current use of each component. Next ask yourself what you would like your students to know, value, and be able to do at the end of your program in relation to their understanding of major biological concepts, unifying principles of biology, processes of scientific investigation, and biological literacy. Finally, determine how you are going to reach your goals, being realistic in the number of changes you are willing to make immediately and in the near future.

Implementing Change

The following recommendations are important to consider as you implement your new curriculum. First, identify the important components of the innovation. This requires an understanding of the content and pedagogy of the new activities you are including in your program. Second, identify leaders of the change. At the high school level, the principal is the agent for change (Fullan, 1982; Leithwood and Montgomery, 1982; and Rutherford, 1986) Successfully implemented programs at this level have had a principal who provided leadership, developed a plan for change, established a change-facilitation team, and supported teachers' efforts for change. At the college level, individual faculty may be more autonomous,

Program Component	Current Practices	Long-term Changes in Practices	How Changes Will Be Made—Short-term and Long-term Actions
Goals for Students and Programs (See Chapters 3 & 4)	Students master content knowledge presented to them at 60% level.	Students become structurally and multidimensionally literate in biology.	Move away from lecture and objective examinations.
Unifying Principles (See Chapter 5)	Don't use.	Organize biological content around six principles.	Rewrite introductory activities to introduce unifying principles.
Biological Knowledge (See Chapter 5)	Determine from list of topics on syllabus or chapters assigned.	Focus on 20 major biological concepts.	Eliminate topics discussed in lecture and replace with activities that introduce and reinforce major biological concepts.
Curriculum Themes (See Chapter 4)	Don't use one.	Organize program around science as a process theme.	Add laboratory component.
Instructional Strategies	See Chapter 4.		
Lecture	Mostly lecture on biological content.	Develop students' ability to use processes of scientific inquiry.	Reduce number of lectures and replace them with discussion periods and labs.
Laboratory	None.	Add investigative experience.	Develop a list of possible future labs.
Educational Technology	None used.	Incorporate interactive technologies.	Review copies of videodiscs available that deal with biology.
Assessment Strategies (See Chapter 4)	Multiple choice exams used.	Provide varied assessment that is appropriate for my goals.	Add written reports on laboratory investigations and independent research project for students this year.
Developing Biological Literacy (See Chapter 3)	Students are at the functional level.	Students have opportunities to practice personal and social decision making.	Search for local issues that students may study on long-term basis. Ask students to evaluate articles presenting scientific information.
Personal Needs of Students	Not addressed in class.	Focus on critical thinking and problem solving skills.	Read references on critical thinking exercises and incorporate a few into class.

Figure 6-1: Example of how you might evaluate your current program

but they still require support of their department chairs and support staff. Third, support and train a change-facilitation team so that each member understands the content and pedagogy of the curriculum. Other biology instructors should be part of this team. Fourth, assist the team in developing an action plan for implementation that includes a description of the ideally implemented curriculum, annual goals, and specific, measurable objectives. Finally, for a new curriculum to be successful, support among instructors is essential both during and after implementation. Individual faculty members should seek out others at neighboring institutions to share concerns, ideas, and materials.

Have I Reached My Goals?

How does an instructor determine whether his or her efforts have been successful? Determination of success, of course, depends on the goals originally established for the program and students. Assessment of students should reflect the diverse learning and teaching strategies used in the classroom, and each assessment strategy should contribute to developing higher levels of biological literacy in your students. One measure of success is the number of students who move from a nominal or functional level of biological literacy to a structural or multidimensional level. Assessment instruments described in Chapter 4 will help determine whether students have attained these levels.

The other part of evaluation is looking at the effectiveness of the entire biology program. In such summative evaluations, information is gathered and analyzed that will be used for course-improvement decisions. This kind of evaluation focuses on the extent to which a course has achieved its stated goals or objectives for the program and learning outcomes for the students—those things learned by students during, or as a part of, the course (Posner and Rudnitsky, 1982). What do your students know and value, and what skills can they demonstrate at the end of your class? We refer you to Chapters 3 and 4 to help you redefine and refine your goals and learning outcomes.

One way to evaluate the effectiveness of a biology program is to observe the students after they have left the classroom. Unfortunately, few teachers have the opportunity to do this. Students may return occasionally to thank you for your course, and you may be able to follow the progress of a few students who move on to undergraduate or graduate school, especially if you have served as their advisor. Of course, the ultimate test of whether your instruction has been successful is whether your students become active, productive, concerned, and informed citizens who are able to use effectively the knowledge and skills that they obtained in your class.

It is important to realize that any changes you make on paper may take time to implement. Typically between three and five years are necessary to integrate an innovation into a curriculum at the high school level (BSCS and IBM, 1989). This process is not necessarily any faster at the collegiate level. As with evolution, change may come about abruptly or over a long period of time. Unlike evolution, however, you have the opportunity to direct the changes toward specific goals.

Conclusion

Any organizational change requires that people change. At the high school level, that means principals, teachers, and support staff must change their attitudes and practices. At the college level, department chairs and support staff must be supportive and cooperative for changes to be implemented. Finally, organizational change requires leadership, and the initiative for change can come from you.

Because biological research generates so much new information each year using revised and new techniques, because many environmental, health, and biological problems remain unsolved, and because students change from year to year, biology educators and their programs must adapt constantly. Thus, you can measure the success of your biology program by your willingness to change, the quality of the goals you set, and the degree to which you attain those goals as reflected in your students.

7

Developing Biological Literacy: A Conclusion

The title—*Developing Biological Literacy*—expresses the major objective of biology education. One aspect of biological literacy involves helping individuals to understand the natural world to fulfill their unique needs for personal development. A second aspect of biological literacy extends to the role individuals have as citizens—to contribute to society. These two objectives need not be separate in biology programs or teaching; rather, they unite and become evident to students as their understanding, appreciation, and skills develop through interactions with the natural world and biology-related personal and social issues.

In this guide, we present biological literacy as more than a slogan. Certainly, the term has the power to rally individuals to the key ideas, values, and skills of educational reform, and it gives biology educators a common language and an objective on which to agree. We formulated concrete examples of biological literacy in the form of unifying principles of biology, scientific processes of inquiry, and scientific and social contexts for the design of biology education programs; thus, we suggest that students' understanding and experiences with these constitute the foundation of biological literacy.

We recognize that the development of biological literacy involves more than increasing students' scientific vocabulary. Although expanding students' vocabulary to include biological terms is part of any biology program, the development of biological literacy combines an understanding of the major biological principles with the processes of science to form a level of biological literacy much broader and deeper than a functional level suggested by the acquisition of vocabulary. Finally, we propose a multidimensional level of biological literacy that adds the richness of history, the nature of biological inquiry, and contemporary advances and issues in biology, biotechnology, and other scientific and mathematical disciplines to our perspective of biological literacy.

In education, the contemporary use of terms such as "scientific literacy" or "biological literacy" implies a general education orientation for high school, community college, and four-year college nonmajors biology programs. In general education courses the primary focus is on the knowledge, values, and skills that students should develop as future citizens, as opposed to future scientists.

Our recognition and reassertion of the general education orientation supports a tradition that includes *General Education in a Free Society* (Harvard Committee, 1945), *General Education in Science* (Cohen & Watson, 1952), and *Science and Liberal Education* (Glass, 1959). Although the philosophical view of *Developing Biological Literacy* is consistent with reports such as those just cited, the practical applications are different. We live in an age in which students experience a broader range of science-related societal issues, have greater access to media, and speak with a louder voice. As biology educators, we have a richer variety of curricular, instructional, and assessment strategies than we had ten, twenty, or a hundred years ago. The teaching of biology by lecture and reading assignments while ignoring the potential of field studies, laboratory experiences, and educational technology constitutes an educational tragedy and ultimately an insupportable approach to educating students who enroll in our biology courses.

Finally, we have tried to provide the background, justification, and information you will need to revise your biology program. We have presented a process that you can use to redesign and implement a new biology program that represents a contemporary orientation and best suits your teaching style, your students, and your particular school. Now, the next, critical step in helping students develop biological literacy is where it ought to be, in your power as a professional biology educator.

References

American Association for the Advancement of Science. 1989. *Science for all Americans: A Project 2061 report on literacy goals in science, mathematics, and technology.* Washington, DC.

___. 1990. High school science textbooks. *Science Books & Films, 25*(4), 171.

Andrews, T. 1964. Undergraduate education of biologists. *BioScience, 14*(13), 15-19.

Association of American Colleges. 1982. *Science and technology education for civic and professional life: The undergraduate years.* A Report of the Wingspread Conference. Washington, DC.

Baron, J. 1991. Performance assessment: Blurring the edges of assessment, curriculum, and instruction. In G. Kulm and S.M. Malcom (Eds.). *Science assessment in the service of reform* (pp. 247-266). Washington, DC: American Association for the Advancement of Science.

Bennett, R.E., W.C. Ward, D.A. Rock, and C. LaHart. 1990. *Toward a framework for constructed-response items* (ETS research report RR 90-7). Princeton, NJ: Educational Testing Service.

Blystone, R.V. 1990. *Learning resource issues in biology.* A commissioned paper for the U.S. Department of Education FIPSE/FIRST Partnership Conference, 17-18 September. Washington, DC.

Biological Sciences Curriculum Study (BSCS). 1963. *Biology teacher's handbook.* New York: John Wiley and Sons, Inc.

___. 1978. An update: The human sciences program. *Biological Sciences Curriculum Study Journal, 1*(2), 5.

___. 1989. *New designs for elementary school science and health.* (A design study supported by International Business Machines.) Dubuque, IA: Kendall/Hunt Publishing Company.

Bybee, R.W. and G. DeBoer. 1993. Goals for the science curriculum. *Handbook of research on science teaching and learning.* Washington, DC: National Science Teachers Association.

Carrier, S.C. and D. Davis-Van Atta. 1987. *Maintaining America's scientific productivity: The necessity of the liberal arts colleges.* Oberlin, OH: Oberlin College Office of the Provost and Instructional Research.

Carter, J.L. (Chair). 1990. *In-depth report on the state of the biology major.* Association of American Colleges task force for liberal education study.

Clark, M. 1989. *Biological and health sciences: A project 2061 report.* Washington, DC: American Association for the Advancement of Science.

Cohen, W. and F. Watson. 1952. *General education in science.* Cambridge, MA: Harvard University Press.

Collins, A. 1992. Portfolios for science education: Issues in purpose, structure, and authenticity. *Science Education. 76*(4), 451-463.

Commission for Undergraduate Education in Biological Sciences (CUEBS). 1967. *Content of core curricula in biology.* [CUEBS publication no. 18]. Washington, DC.

Council of Chief State School Officers (CCSSO). 1990. *State indicators of science and math 1990.* Washington, DC: CCSSO.

Demastes, S. and J.H. Wandersee. 1992. Biological literacy in a college biology classroom. *BioScience, 42*(1), 63-65.

Educational Policies Commission. 1966. *Education and the spirit of science*. Washington, DC: National Education Association.

Ewing, M.S., N.J. Campbell, and M.J.M. Brown. 1987. Improving student attitudes toward biology by encouraging scientific literacy. *The American Biology Teacher, 49*(6), 348-350.

Fullan, M. 1982. *The meaning of educational change*. New York, NY: Teachers College Press, Columbia University.

Gibbs, A. and A.E. Lawson. 1992. The nature of scientific thinking as reflected by the work of biologists and by biology textbooks. *The American Biology Teacher*, 54(3), 137-151.

Glass, B. 1959. *Science and liberal education*. Baton Rouge, LA: Louisiana State University Press.

Haertel, E. 1991. Form and function in assessing science education. In G. Kulm and S.M. Malcom (Eds.). *Science assessment in the service of reform* (pp. 233-245). Washington, DC: American Association for the Advancement of Science.

Harvard Committee. 1945. *General education in a free society*. Cambridge, MA: Harvard University Press.

Hawkins, J. and K. Sheingold. 1986. The beginning of a story: Computers and organization of learning in classrooms. In J.A. Culbertson and L.L. Cunningham (Eds.), *Microcomputers and education, 85th yearbook of the NSSE.* (pp. 40-58). Chicago, IL: The National Society for the Study of Education.

Hein, G. 1991. Active assessment for active science. In V. Perrone (Ed.), *Expanding student assessment*, (pp. 106-131). Alexandria, VA. Association for Supervision and Curriculum Development.

Hudson, L. 1991. National initiatives for assessing science education. In G. Kulm and S.M. Malcom (Eds.). *Science assessment in the service of reform* (pp. 107-126). Washington, DC: American Association for the Advancement of Science.

Hurd, P.D. 1961. *Biology education in American schools: 1890-1960.* (Biological Sciences Curriculum Study Bulletin No. 1). Washington, DC: American Institute of Biological Sciences.

Hurd, P.D., R.W. Bybee, J.B. Kahle, and R. Yager. 1980. Biology education in secondary schools of the United States. *The American Biology Teacher, 42*(7), 388-410.

Johnson, R.C. 1984. Science, technology, and black community development. *The Black Scholar, 15*(2), 32-44.

Johnson, D. and R. Johnson. 1991. Group assessment as an aid to science instruction. In G. Kulm and S.M. Malcom (Eds.), *Science assessment in the service of reform* (pp. 283-289). Washington, DC: American Association for the Advancement of Science.

Jones, G. 1989. Biological literacy. *The American Biology Teacher, 51*(8), 480-481.

Kahle, J.B. 1990. *Issues in instruction: Biology.* A commissioned paper for the U.S. Department of Education FIPSE/FIRST Partnership Conference, 17-18 September. Washington, DC.

Leithwood, K.A. and D. Montgomery. 1982. The role of the elementary school principal in program improvement. *Review of Educational Research*, 309-339.

Lester, F. and D. Kroll. 1990. Assessing student growth in mathematical problem solving. In G. Kulm and S.M. Malcom (Eds.), *Assessing higher order thinking in mathematics* (pp. 53-69). Washington, DC: American Association for the Advancement of Science.

Linn, M.C. 1987. Designing science curricula for the information age. Commissioned paper for the IBM-supported *Design Project*. Colorado Springs, CO: Biological Sciences Curriculum Study.

Livermore, A.H. 1964. The process approach of the AAAS Commission on Science Education. *Journal of Research in Science Teaching, 2*, 271-282.

Millman, J. and J. Greene. 1989. The specification and development of tests of achievement and ability. In R.L. Linn (Ed.). *Educational measurement, 3rd edition* (pp. 335-366). New York, NY: Macmillan.

Mokros, J.R. and R. Tinker. 1986. *The impact of microcomputer-based science labs on children's graphing skills*. A paper presented at the annual meeting of the National Association of Research in Science Teaching, San Francisco, CA.

Moore, J.A. 1983. *Science as a Way of Knowing. Volume I: Evolutionary biology*. Baltimore, MD: American Society of Zoologists.

____. 1984. Science as a way of knowing. Evolutionary biology. *American Zoologist 24*(2), 470.

____. 1984. *Science as a Way of Knowing. Volume II: Human ecology*. Baltimore, MD: American Society of Zoologists.

____. 1985. *Science as a Way of Knowing. Volume III: Genetics*. Baltimore, MD: American Society of Zoologists.

____. 1986. *Science as a Way of Knowing. Volume IV: Developmental biology*. Baltimore, MD: American Society of Zoologists.

____. 1987. *Science as a Way of Knowing. Volume V: Form and function*. Baltimore, MD: American Society of Zoologists.

____. 1988. *Science as a Way of Knowing. Volume VI: Cell and molecular biology*. Baltimore, MD: American Society of Zoologists.

____. 1989. *Science as a Way of Knowing. Volume VII: Neurobiology and behavior. (Part 1)* Baltimore, MD: American Society of Zoologists.

____. 1990. *Science as a Way of Knowing. Volume VII: Neurobiology and behavior. (Part 2)* Baltimore, MD: American Society of Zoologists.

National Assessment of Educational Progress. 1987. *Learning by doing*. Princeton, NJ: Educational Testing Service.

National Commission on Excellence in Education. 1983. *A nation at risk: The imperative for educational reform*. Washington, DC: U.S. Department of Education.

National Research Council. 1982. *Science for non-specialists: The college years*. Washington, DC: National Academy Press.

____. 1990. *Fulfilling the promise: Biology education in the nation's schools*. Washington, DC: National Academy Press.

National Science Board. 1986. *Undergraduate science, mathematics, and engineering education*. Washington, DC.

National Science Foundation. 1989. Report on the National Science Foundation disciplinary workshops on undergraduate education. Washington, DC.

____. 1989. *Report on the National Science Foundation workshop on science, engineering, and mathematics education in two-year colleges*. Washington, DC.

National Science Teachers Association. 1991. Science-technology-society: A new effort for providing appropriate science for all. *NSTA Reports*, 36-37.

O'Neil, J. 1992. Schools try problem-based approach. *Wingspread: The Journal, 14*(3), 1-3.

Papert, S. 1981. Computers and computer cultures. *Creative Computing, 7*(3), 82-92.

Pea, R.D. and D.M. Kurland. 1984. On the cognitive effects of learning computer programming. *New Ideas in Psychology, 2*, 137-168.

Perrone, V. 1991. *Expanding student assessment*. Alexandria, VA: Association for Supervision and Curriculum Development.

Pickering, M. 1985. Lab is a puzzle, not an illustration. *Journal of Chemical Education*, *62*(10), 874-875.

Posner, G. and A. Rudnitsky. 1982. *Course design: A guide to curriculum development for teachers*. New York, NY: Longman, Inc.

Project Kaleidoscope. 1991. *What works: Building natural science communities*. Volume One. Washington, DC: Stamats Communications, Inc.

Roberts, D.A. 1982. Developing the concept of "curriculum emphases" in science education. *Science Education*, *66*(2), 243-260. John Wiley & Sons, Inc.

Rosen, W.G. 1989. *High school biology: Today and tomorrow*. Washington, DC: National Academy Press.

Rosenthal, D.B. and R.W. Bybee. 1987. Emergence of the biology curriculum: A science of life or a science of living? In T.S. Popkewitz (Ed.), *The formation of the school subjects* (pp. 123-144). Philadelphia, PA: The Falmer Press.

___. 1988. High school biology: The early years. *The American Biology Teacher*, *50*(6), 345-347.

Rutherford, W.L. 1986. School principals as effective leaders. *Phi Delta Kappan*, *69*, 31-34.

Sigma Xi, The Scientific Research Society. 1989. *An exploration of the nature and quality of undergraduate education in science, mathematics and engineering: A report of the National Advisory Group*. New Haven, CT.

Stiggins, R.J. and N.J. Bridgeford. 1986. Student evaluation. In R.A. Berk (Ed.). *Performance assessment: Methods & applications*. (pp. 469-491). Baltimore, MD: The Johns Hopkins University Press.

Tinker, R. 1984. *Microcomputer-based laboratories*. Cambridge, MA: Technical Education Research Centers.

Uno, G. 1990. Inquiry in the classroom. *BioScience*. *40*(11), 841-843.

U.S. Congress, Office of Technology Assessment. 1992. *Testing in American schools: Asking the right questions*. (OTA-SET-519). Washington, DC: U.S. Government Printing Office.

Vetter, B. 1989. Look who's coming to school. Changing demographics: Implications for science education. In *Curriculum Development for the Year 2000*, (pp. 1-12), Colorado Springs, CO: Biological Sciences Curriculum Study (BSCS).

Walker, D.B. 1990. *The biology curriculum: A call for radical adaptation*. A commissioned paper for the U.S. Department of Education FIPSE/ FIRST Partnership Conference, 17-18 September. Washington, DC.

Wandersee, J.H. 1990. Using the past to predict the future of bioeducation. *The American Biology Teacher*, *52*(6), 346-351.

Weiss, I. 1978. *Report of the 1977 national survey of science, mathematics, and social studies education*. Washington, DC: U.S. Government Printing Office.

___. 1987. *Report of the 1985-86 national survey of science and mathematics education*. Research Triangle Park, NC: Center for Educational Research and Evaluation.

Wittrock, M.C. and E.L. Baker. 1991. *Testing and cognition*. Englewood Cliffs, NJ: Prentice Hall.

Wolf, D.P. 1989. Portfolio assessment: Sampling student work. *Educational Leadership*. *46*(7), 35-40.

Wright, E.L. and R.K. Backe. 1992. Towards excellence in biology teaching. *The Science Teacher*, *59*(5), 32-35.

Wright, E.L. and G. Govindarajan. 1992. A vision of biology education for the 21st century. *The American Biology Teacher*, *54*(5), 269-272.

Yager, R. and P. Tweed. 1991. Planning a more appropriate biology education for schools. *The American Biology Teacher*, *53*(8), 479-483.

Unifying Principles of Biology

Evolution: Patterns & Products of Change

Interaction & Interdependence

Genetic Continuity & Reproduction

Growth, Development, & Differentiation

Energy, Matter, & Organization

Maintenance of a Dynamic Equilibrium

Biological Knowledge

is

Unified by EVOLUTION

and is related to different

Unifying Principles,

Levels of Organization,

and

Living Systems

Biology is an Active Process

Observing & Comparing

Asking Questions & Identifying Problems

Forming Hypotheses & Predicting Outcomes

Designing & Conducting Experiments

Collecting & Organizing Data

Analyzing Data & Relating Ideas

Proposing Explanations & Making Inferences

Communicating Explanations
& Applying Knowledge

Assembly Directions

- Cut along the outer edge of the figure on this sheet and the accompanying sheet.
- Fold along bold lines.
- Arrange cube so that the two BSCS copyrights are on opposite sides.
- Place glue on top side of tabs and stick each tab to the adjoining side.

Curriculum Themes

Science as a Process

Unifying Principles of Biology

Biological Content

Personal Aspects of Biology

Social Aspects of Biology

Ethical Aspects of Biology

Technology

Historical Development
of Biological Ideas

Learning/Teaching/Assessment Strategies

Lab Activities / Field Experiences

Inquiry and Problem Solving

Individual Research

Discussion / Debates

Cooperative Learning

Students Read / Write / Speak / Explain

Computer Interactions / Educ Tech

Testing Hypotheses / Analyzing Data

Performance / Authentic / Self Assessment

Research, Project, Lab, & Oral Reports

Written Exams / Lab Practicals

Observe Student Behaviors / Interviews

Portfolios / Models & Simulations

Biological Literacy

A biologically literate person should:

Understand

Biological Principles

Impact of Humans on Biosphere

Scientific Inquiry

Historical Development

Value

Scientific Investigations

Biodiversity & Cultural Diversity

Biology & Biotechnology in Society

Importance of Biology to the
Individual

Be Able To

Think Creatively

Reason Logically & Critically

Use Technologies Appropriately

Make Personal & Social Decisions

Apply Knowledge to Solve Problems